建筑立场系列丛书 No.66

创意办公

The Creative Workplace

让·努维尔建筑师事务所等 | 编

孙探春 于风军 杜丹 王京 高莹 陈帅甫 | 译

大连理工大学出版社

C3 建筑立场系列丛书 No. 66

创意办公

知识经济时代的办公室设计

The Creative Workplace

Office Design for the Knowledge Economy

垃圾焚烧发电厂 _Schmidt Hammer Lassen Architects+Gottlieb Paludan Architects

丹麦施密特–哈默–拉森建筑师事务所的团队与戈特利布·帕卢丹建筑师事务所赢得了世界上最大的垃圾焚烧发电厂的国际设计竞赛。新建成的深圳东部垃圾焚烧发电厂坐落于深圳市山岭起伏的市郊，每天将焚烧5000吨垃圾，这相当于深圳的2000万市民每年生产垃圾总量的三分之一。该发电厂将采用最先进的垃圾焚烧与发电技术，同时也担负起培养深圳市民环保意识的教育责任。

获胜设计在一座圆形建筑内系统规划了整个发电厂，包含附属建筑，打破了传统工业厂房长方形布局的印象。这种清晰的圆形使发电厂的占地面积最小化，也就减少了现场必要的土方挖掘量。

公众游客先是穿过一座园景公园，再进入发电厂，走过一条入口桥，这座桥架起在前厅门前的两根大烟囱与游客中心之间，俯瞰着发电厂的机械设备。一条内部环形路径与通道环绕着发电厂，向人们说明了每一道工序，最后将人们带到了屋顶一条1.5km长的全景公共通道，在这里可俯瞰周边美景以及深圳市容。

建筑面积为66000m^{2}的屋顶上面覆盖了多达44000m^{2}的太阳能板，因而该发电厂不仅能以更加清洁的方式处理城市垃圾，也能对这座城市的再生能源供应做出贡献。

发电厂有意识地展示了垃圾焚烧发电的工序，将此作为一个重要的技术过程，既适合解决日益增多的垃圾问题，也能找到更加环保的发电方式。与此同时，游客也了解到了人们每天制造越来越多垃圾所带来的挑战，并且得到了教育，愿意积极主动减少他们自己每天制造的垃圾量。

深化设计工作在2016年年初展开，发电厂预期将于2020年投入运行。

Waste-to-Energy Plant

The team of Schmidt Hammer Lassen Architects and Gottlieb Paludan Architects has won an international competition to design the largest waste-to-energy plant in the world. Located on the mountainous outskirts of Shenzhen, the new Shenzhen East Waste-to-Energy Plant will incinerate 5,000 tonnes of waste per day - equaling one third of the waste generated by Shenzhen's 20 million inhabitants every year. The plant will utilize the most advanced technology in waste incineration and power generation, whilst at the same time act as a source of education for the citizens of Shenzhen. The winning design organizes the entire plant, including auxiliary buildings, into one circular building - breaking with the traditional rectangular layout of industrial facilities. By proposing a clear circular form, the footprint of the plant is minimised and it reduces the amount of excavation required to build on the site.

Public visitors are invited into the plant through a landscaped park, via an entrance bridge that rises between the stacks to an entrance lobby and visitor center overlooking the plant machinery. An internal circular path and walkway circle the plant explaining each process, before leading up to a 1.5km panoramic public walkway on the roof overlooking the surrounding landscape and the city of Shenzhen.

The 66,000m² roof is designed to be covered by up to 44,000m² of photovoltaic panels providing the opportunity for the plant not only to provide a cleaner way to deal with the city's waste but also to contribute to the renewable energy provision for the city.

The plant is intended to showcase the waste-to-energy production as an important technical process that is geared to deal with the issues of growing waste, as well as the issue of finding more environmentally friendly ways of generating electricity. At the same time visitors become informed on the challenge of the growing amounts of waste we produce every day and are also educated on initiatives on how to reduce their own amount of daily waste.

Detailed design work is about to begin in early 2016, and the plant is scheduled to start operating in 2020.

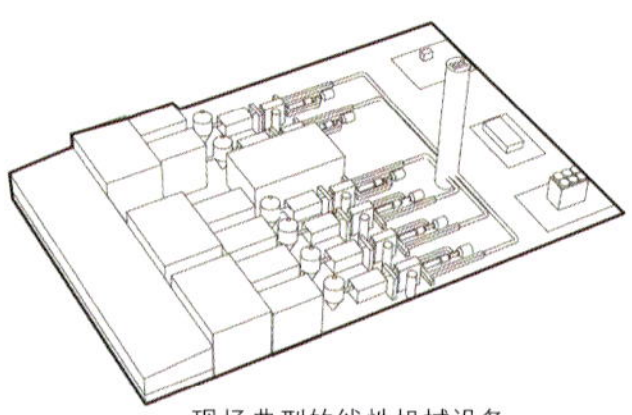

现场典型的线性机械设备
standard linear machinery on site

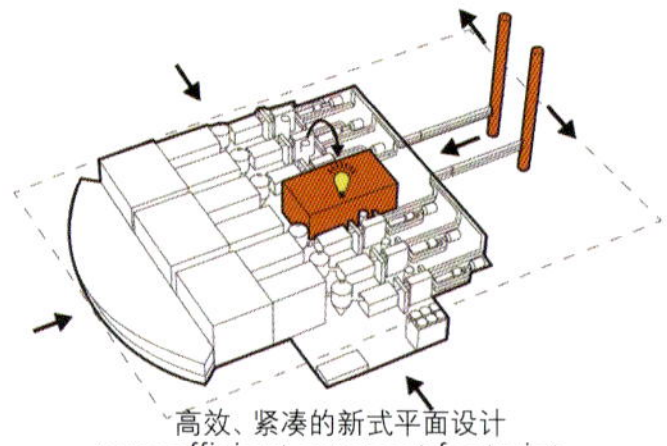

高效、紧凑的新式平面设计
new efficient compact footprint

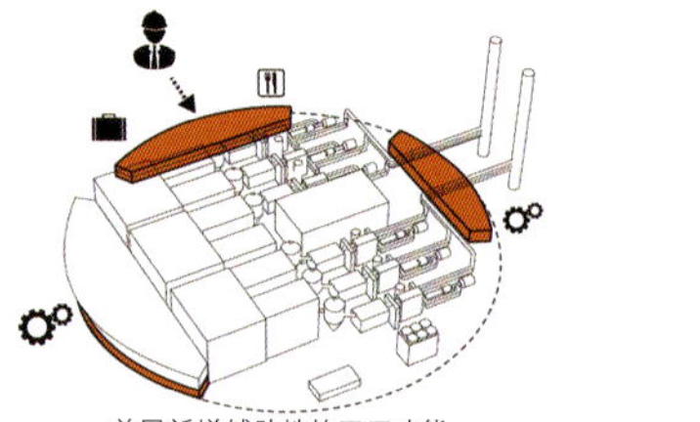

首层新增辅助性的工厂功能
added supporting factory functions on the ground floor

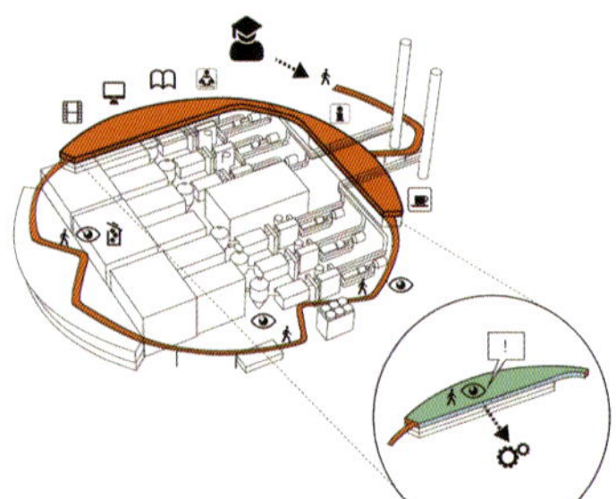

在屋顶新增游客体验
add new visitor experience on top

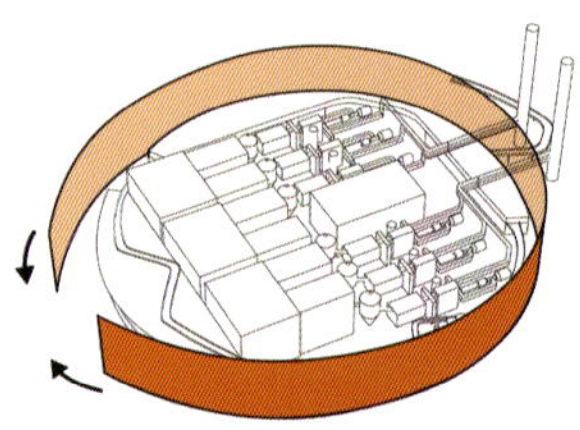

立面外围护
facade wrap

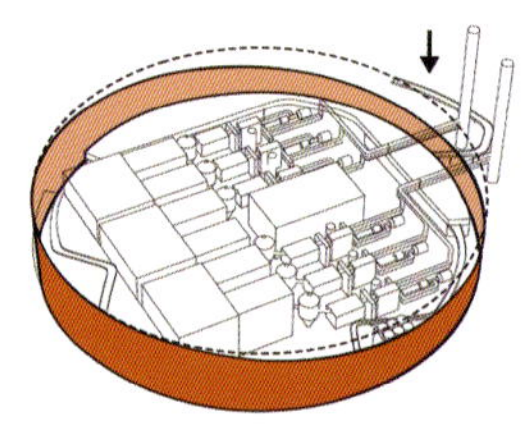

调整高度，优化体量
adjust height
to optimize volume

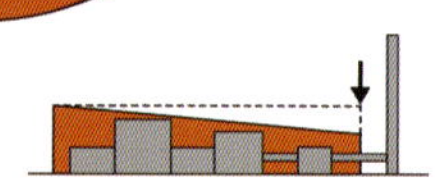

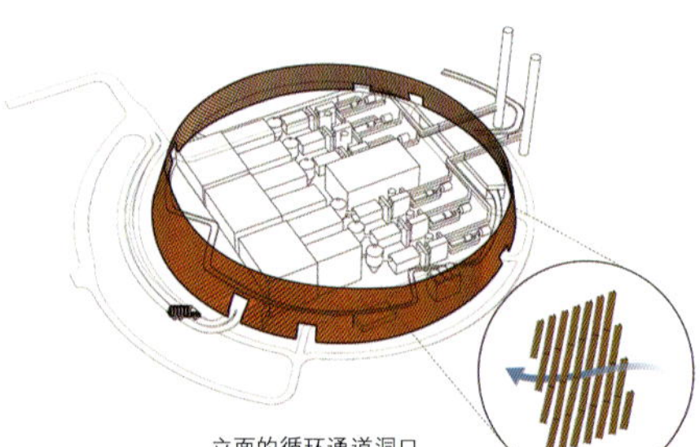

立面的循环通道洞口
facade openings to circulation

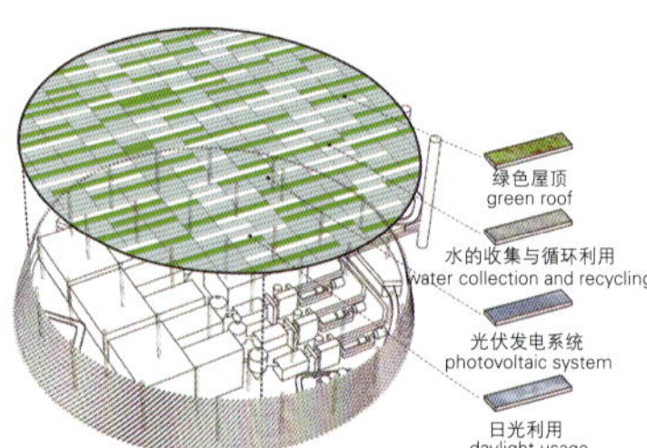

屋顶系统的智能设计
intelligent plug-in roof system

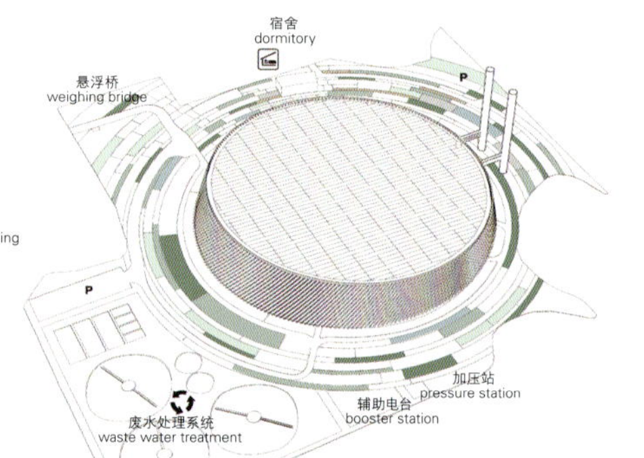

主题景观带在发电厂周边呈辐射状分布
thematic landscape strips radiate
around the plant

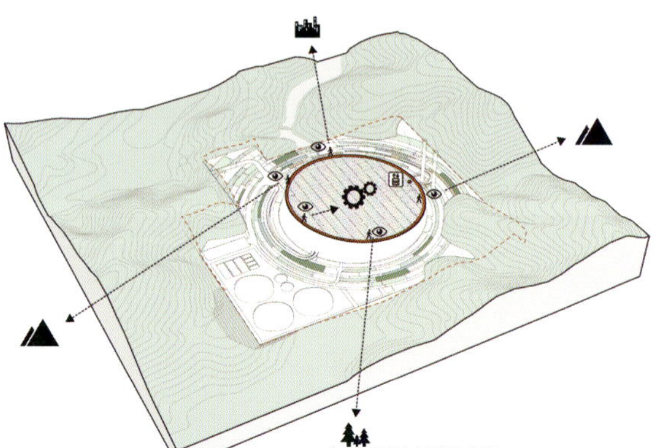

新增屋顶全景发现小路
add rooftop viewing panorama discovery path

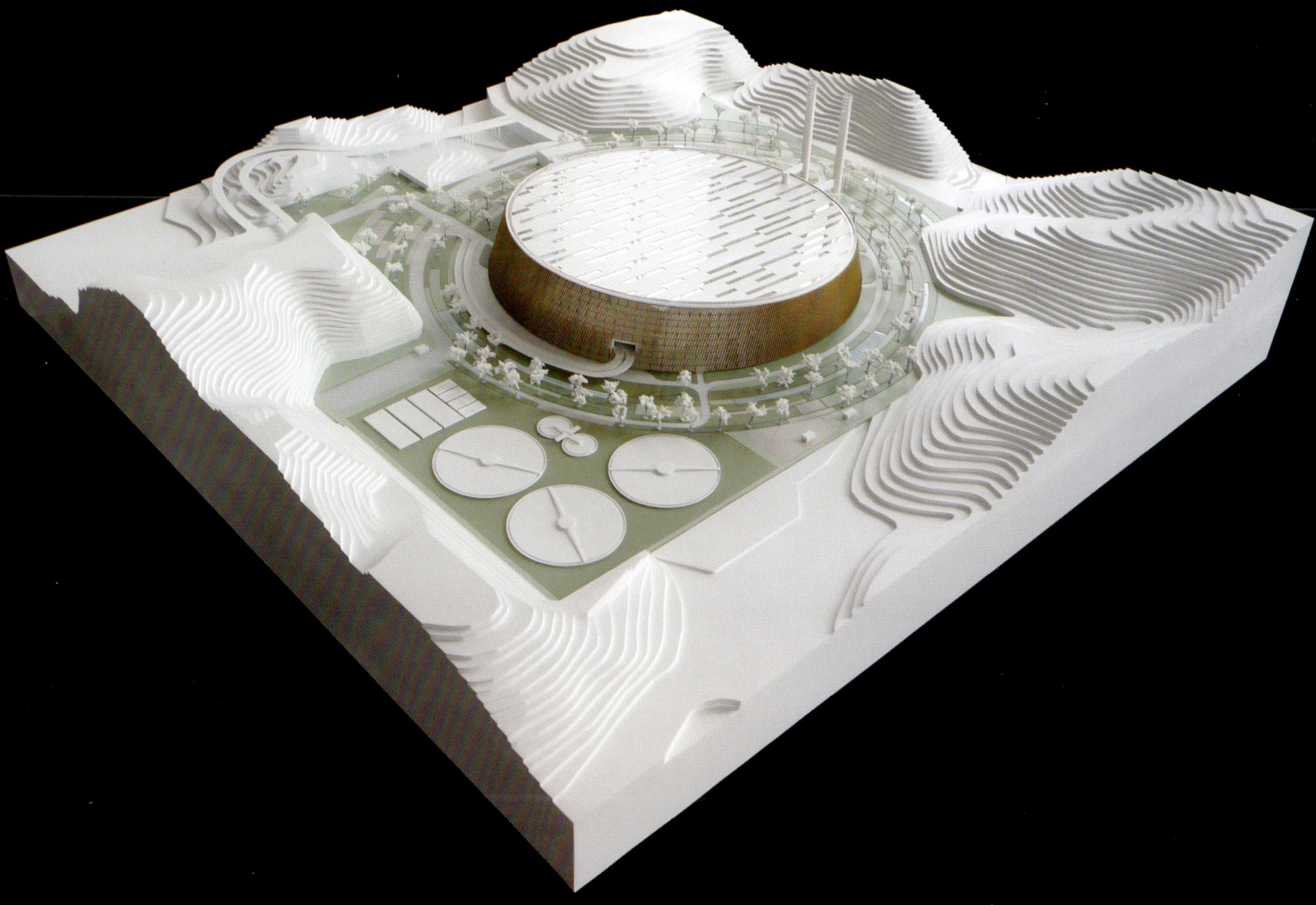

可持续建筑之家 _hiperstudio

本项目是"巴西可持续建筑之家国家建筑设计竞赛"的获胜者。此活动由坎皮纳斯市市政厅发起，由IAB|SP负责组织。项目建于坎皮纳斯市塔夸拉尔公园内即将建造的一座可持续性的机构场馆之内。

身临该项目其境，人们相信，设计这一行为不仅仅要考虑功能和美学这两个方面的问题，还要在建筑形式中体现社会和环境的重要性，需要敏感和人性化的设计方法。

这一项目设计致力于探索可持续性这一问题，以超越人们惯常理解的"可持续性"的意思。人们通常把可持续性建筑与绿色建筑联系在一起，即建筑墙体后面所隐藏的一套程序和技术机制。然而，这一设计的目标远非如此。该项目旨在调整可持续性这一理念中本身所固有的人体尺寸，注重用户体验，即人们真正能感受到的东西，从而最终让人们重新思考在环境中的作用以及如何与环境相处。

该设计方案主要基于一种永恒的建筑理念：将建筑人行漫步道作为项目的一种设计姿态。作为设计理念和策略，人行漫步道从建筑的主入口就呈现在人们眼前。从此处，人们沿着精心设计的小路开始了步行体验，去发现与周围景观融为一体的建筑形式。

在建筑与环境共生共栖的关系中，建筑在人们对周围自然环境沉思冥想的过程中起到了实际支持的作用。场馆结构内包含了各种各样的主题活动，精心设计的小路环绕四周，一直延伸到屋顶。人们可在与树顶齐高的屋顶继续畅游，直到登上更高的平台，饱览塔夸拉尔公园的风景。

因此，小路对使用者而言起着诲人不倦的作用，使人们认识到自然环境和人造环境之间存在着充满诗意的互动。其中，建筑是配角，对主角——公园的景色充满敬畏。

最后，该项目被认为是可持续性建筑的典范，其中包含了一系列能保证环保性能相互协调的系统和设计解决方案。该建筑设计方案达到

了客户关于建筑要具有标志性形象的要求，而且还使用了必要而经济的构件，使建筑简洁而富有表现力，没有形式上的扭曲或者是材料上的奢侈浪费。

House of Sustainability

This project was selected as the winner of a National Architecture Competition for the House of Sustainability, promoted by Campinas City Hall and organized by IAB | SP. It consists in a sustainable institutional pavilion to be built in Taquaral Park, Campinas.

In the premises of this project, there is the belief that the act of designing does not operate only between the functional and aesthetic issues, but also seeks to translate social and environmental significance in the built form, requiring a sensitive and humane approach.

This proposal seeks to explore the issue of sustainability in order to transcend its meaning more commonly associated with green architecture, i.e., a mere set of procedures and technological mechanisms hidden behind the walls. More than that, the project aims to rescue the human dimension intrinsic to the very notion of sustainability, focused on the user experience, something he can actually feel and sense,

and consequently, rethink his role and his way of relating to the environment.

The main idea behind the design proposal is based on one of the timeless concepts of architecture: the architectural promenade as a designing gesture of the project. The promenade as a conceptual strategy is evidenced since the main access of the building site, where the walking experience along the proposed path begins, leading to the discovery of the architectural form integrated with the landscape.

In that symbiotic relationship, the architecture acts as a physical support for the contemplation of the natural surroundings. The proposed path unfolds along the pavilion that contains all the required programmatic activities, rising up to the roof where the tour continues at the height of the treetops, culminating in a higher terrace that allows privileged views of Taquaral Park.

Thus, the path plays a didactic role for the users, raising the awareness towards the poetic interaction between the natural and built environment, in which architecture is resigned to a supporting role in a gesture of reverence to the protagonist, the landscape of the Park.

Lastly, the project is conceived as a paradigmatic prototype of sustainable architecture, containing a series of systems and design solutions that ensure an integrated environmental performance. The architectural proposal responds to the demands of the client about the iconic character, yet with constructive simplicity and expressiveness found in essential and economic elements, without formal contortions or extravagant materiality.

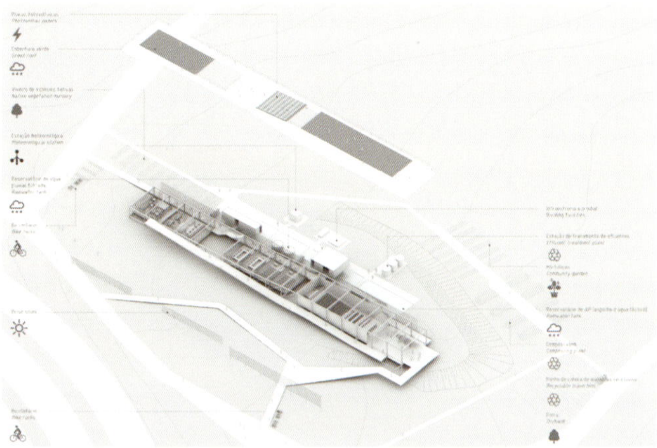

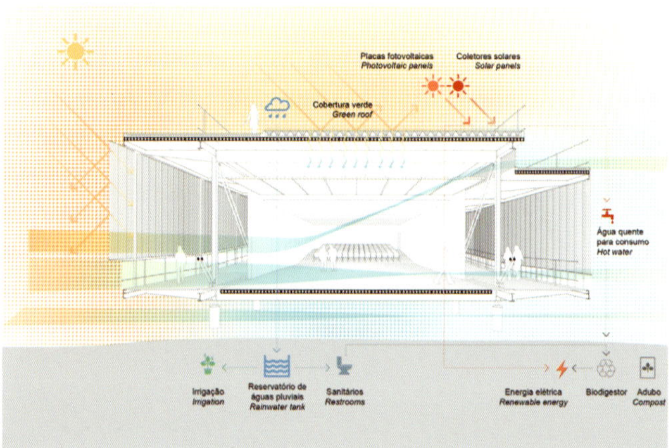

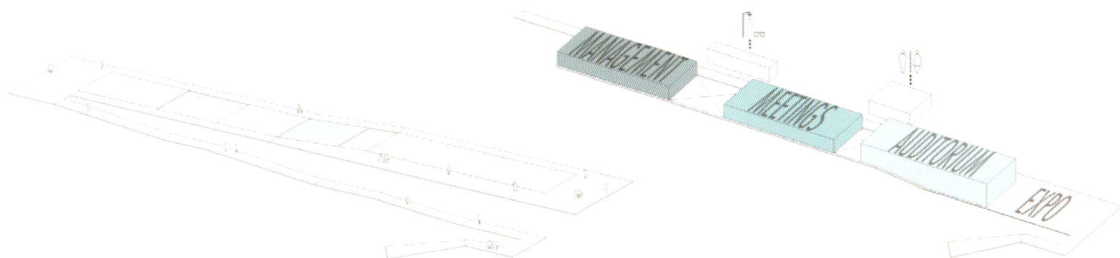

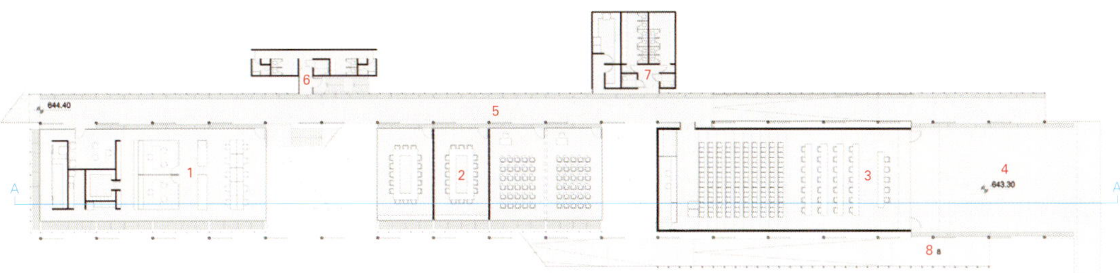

1. 管理办公室 2. 会议中心 3. 礼堂 4. 展览区 5. 循环区域 6. 卫生间/更衣室 7. 公共卫生间/厨房/储藏室 8.门廊
1. management 2. convention center 3. auditorium 4. exhibition area
5. circulation area 6. restrooms/changing rooms 7. public restroom/kitchen/storage 8. porch
一层 ground floor

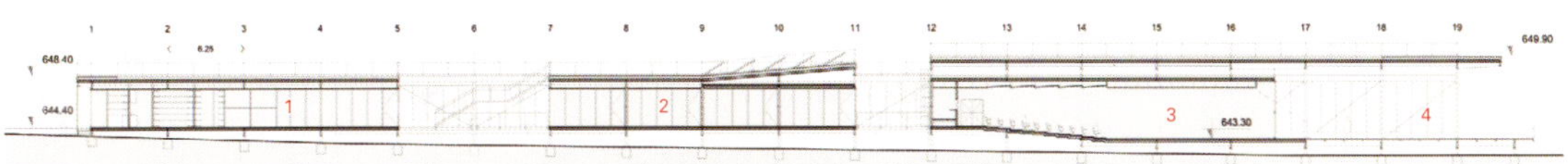

1. 管理办公室 2. 会议中心 3. 礼堂 4. 展览区
1. management 2. convention center 3. auditorium 4. exhibition area
A-A' 剖面图 section A-A'

加拿大独木舟博物馆_Heneghan Peng Architects

建筑师遴选委员会的委员和加拿大独木舟博物馆董事会共同宣布赫尼根&彭建筑师事务所（爱尔兰都柏林）和卡恩斯·曼奇尼建筑师事务所（加拿大多伦多）最终赢得了加拿大独木舟博物馆国际设计竞赛大奖。新建的加拿大独木舟博物馆位于1904年修建的彼德堡水闸国家历史遗址上，选址宏伟而庄严，建设经费达3000万欧元。

经过两个阶段的国际竞赛，最后的获胜方案设计了一个优雅、蜿蜒曲折的玻璃亭子，顶部为2英亩（约合8093.7m²）的屋顶花园。建成后，这里将成为世界上最大的独木舟和皮划艇收藏博物馆。本设计令人耳目一新，在特伦特-塞文水道旁的鼓丘开始，蜿蜒曲折，设计做得系统而大胆。博物馆设计心系社区，为了社区，充分体现了土著居民的智慧：在大地上轻松地生活，轻松地（不过度地）建造。

赫尼根&彭建筑师事务所和卡恩斯·曼奇尼建筑师事务所的设计方案从其他设计方案中脱颖而出，因为他们的设计有机地与当地的环境融为一体，而非凌驾于周围环境之上。处于当今气候变化的时代，该设计打动设计评审委员会的不仅仅是其地热供暖/制冷系统设计以及降低能源成本的方式，而且设计中体现的可持续发展方面的智慧也给评委留下了深刻的印象。在这一嵌入式的设计中，只有东面和南面的玻璃墙暴露在外，因而设计本身就降低了运营成本。博物馆只有一层，内部面积有7500m²，平面布局可以非常灵活，为观众带来各种不同的体验和服务，适应各种变化的预期改造需要和不断发展的技术要求。

这种有机造型的建筑体量嵌入鼓丘，顶部边缘使用当地硬木铺设，如同一条飘逸的丝带。这样，博物馆中对光敏感的藏品，如可以追溯到18世纪80年代、具有历史意义的桦树皮独木舟以及土著手工艺品，就可以放置在被动式节能的自然避光的地方。博物馆的大面积绿

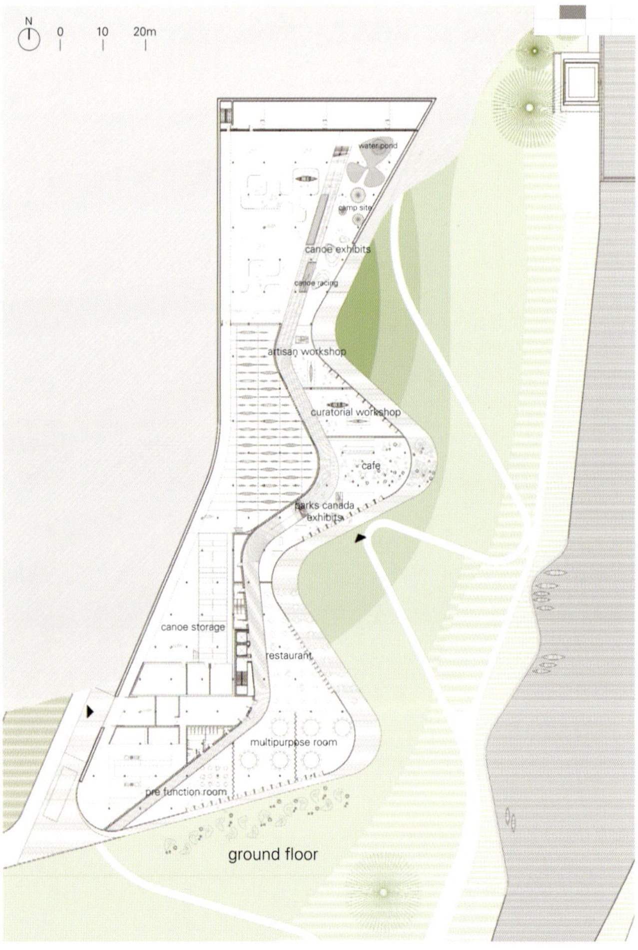

色屋顶可以成为社区种植园，种植可食用的植物，可以吸引当地的花粉传播者，可以践行原住民的三姐妹种植作物（玉米、南瓜、豆子）的种植。与此同时，可以对暴雨的雨水进行有效管理，也可以在此饱览水闸的旖旎风光。

"评审委员会判定赫尼根&彭建筑师事务所/卡恩斯·曼奇尼建筑师事务所获得此次竞赛大奖是因为他们的设计彰显了独木舟的轻盈及其绝妙超群的功能。该博物馆的设计打破了以自我为中心的建筑理念，创造了一个温文尔雅、天然有机的空间，与特伦特-塞文水道蜿蜒相伴，如诗如画。无论是身处博物馆内部，还是站在博物馆屋顶，你总是能看见水，看见整装待发的独木舟。我已迫不及待地想看到这所博物馆竣工了。它将会改变我们对建筑的看法，对如何营造一个场所的看法以及对独木舟的看法：这是一个真正的标志性设计。"

Canadian Canoe Museum

Members of the Architect Selection Committee and the Board of Directors of the Canadian Canoe Museum announce that Heneghan Peng Architects (Dublin, Ireland) and Kearns Mancini Architects (Toronto, Ontario) have won the international competition for the new €30 million Canadian Canoe Museum to be located at the majestic site of the 1904 Peterborough Lift Lock National Historic Site.

An elegant, serpentine glass pavilion graced by a two-acre rooftop garden has been selected as the winning design in the two-stage international competition. The design to house the world's largest collection of canoes and kayaks presents a Canadian game changer that organically and boldly curves out from the drumlins beside the Trent-Severn waterway. Envisioned with and for the community, the museum embraces

aboriginal wisdom to live and build lightly on the land.
The Heneghan Peng Architects/Kearns Mancini submission stood apart from the other submissions as the design works organically with the land rather than overwhelming it. In an era of climate change, its intelligence on sustainability impressed the design jury, not only for its geothermal heating/cooling and reduced energy costs. The embedded design has inherently lower operating costs with only the east and south glass walls exposed to the elements. Inside, the 7500 square metre single floor design offers a flexible floor plate, allowing the museum to reconfigure the experience and offerings, adapting to changing expectations and technology.
The organically-shaped volume banded on its top edge with local hardwood is embedded within the site's drumlins, allowing the museum's light-sensitive collections of historic birch bark canoes that date back to the 1780s and aboriginal artefacts to enjoy energy-passive, naturally dark spaces. The museum's two-acre green roof will provide the community with the possibility of creating edible gardens, native flower pollinators and aboriginal three sister plantings while facilitating efficient management of storm water and fantastic views to the Lift Lock.
"The jury awarded the Heneghan Peng Architects/Kearns Mancini scheme for exemplifying the lightness and sublime functionality of the canoe. The museum design breaks with ego-driven architecture to offer a gentle, organic space that poetically winds its way along the Trent-Severn. No matter where you are inside the museum or outside standing on its roof, you will always be able to see the water, and a canoe waiting to be paddled by you. I can't wait to see this museum built. It's going to change the way we think about architecture, place-making and the canoe – a true icon of design."
Lisa Rochon, Chair, Canadian Canoe Museum Selection Committee

伊斯塔帕拉帕的帕帕洛特儿童博物馆 _MX_SI+SPRB Arquitectos

MX_SI和SPRB建筑师事务所合作设计的方案赢得了位于墨西哥城东部的伊斯塔帕拉帕的帕帕洛特儿童博物馆设计竞赛。该项目设计方案的口号是"让我们打造城市"，清楚表达了使该项目成为都市环境中的活化剂的设计意图。这是一个综合性博物馆设计，面向城市开放，与公共空间融为一体，建筑体量既在地面层与整个城市保持实际的连续性，又在上层空间通过通透的立面开窗设计与整个城市保持视觉上的连续性。

博物馆的城市规划策略是在街道层面上恢复公共空间作为聚会场所的价值，同时与其所在地的城市节点和谐地呼应。该建筑从街道上撤后几米，形成一个城市广场，既是博物馆的延伸，同时又相当于宽敞明亮的公共大厅。一面面柱墙犹如一片森林，人们穿过这一模糊的边界进入博物馆的中心，因而将广场引入博物馆内部。

咖啡馆、礼品店等公共区域位于一层，人们可以随意出入；展览区域位于夹层以上几层，采用交互式的空间布局，多功能室、IMAX影院和剧场的休息厅就位于夹层。博物馆不同的楼层有不一样的主题，但每个公共区域都有一个像露台和花园这样的室外空间，都有各自不同的特点。

博物馆的楼层平面图按照10m宽的长条状布局，呼应了建筑模块的结构规律，由裸露的混凝土隔墙组成，隔墙再与V形房梁连为一体。这些隔墙构成了抽象意义上的森林，其纹理结构经过精心设计，参照了图莱树（一种巨大的墨西哥柏树）的树干，底部结实、坚固，向上长出枝条，越往上越窄。建筑模块根据水平和垂直移动的策略分组，构成了新博物馆的体量。

Papalote Children's Museum in Iztapalapa

MX_SI and SPRB has won a competition to design Papalote Children's Museum in Iztapalapa, in the east of Mexico City. With the slogan "Let's Make City", the project is clearly intended to be an activator in the urban environment. It is a design for an integrative museum that opens up to the city and is incorporated into public space, creating physical continuity on the ground floor and visual continuity on the upper storeys, via great windows facing to the city.

The urban strategy of the museum sets out to recover the value of public space as a meeting place at street level and respond harmoniously to the urban junction in which it is

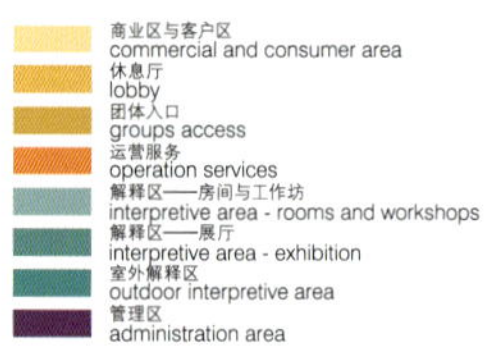

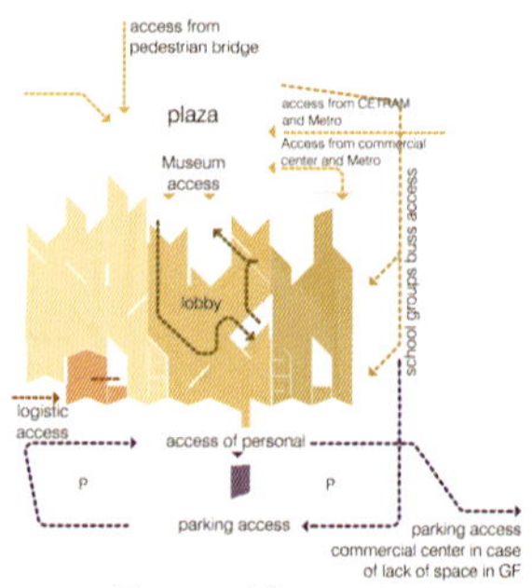

一层 ground floor

夹层 mezzanine floor

二层 first floor

三层 second floor

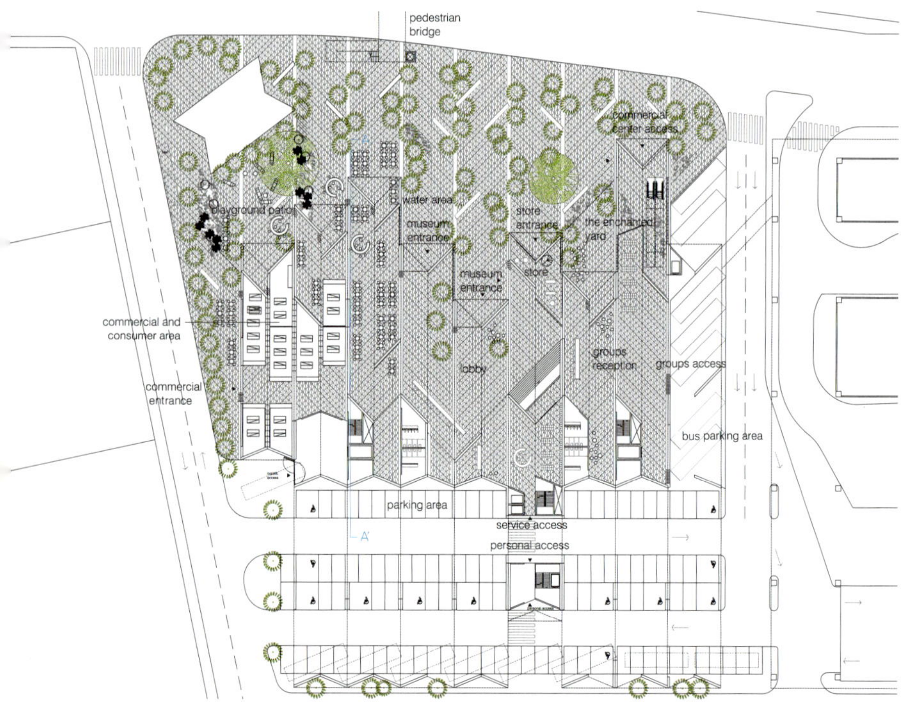

一层 ground floor

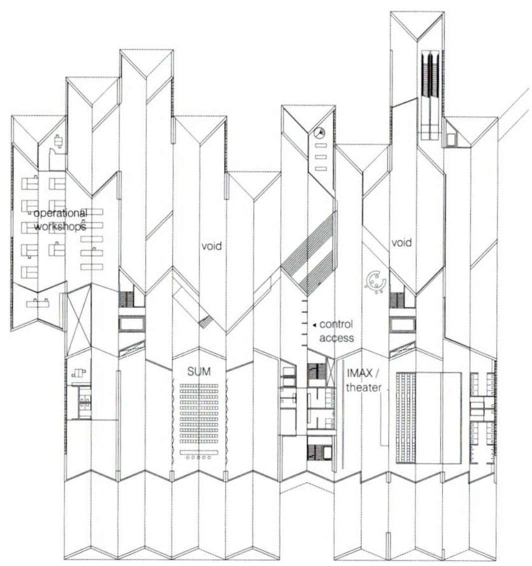

夹层 mezzanine floor

南立面 south elevation

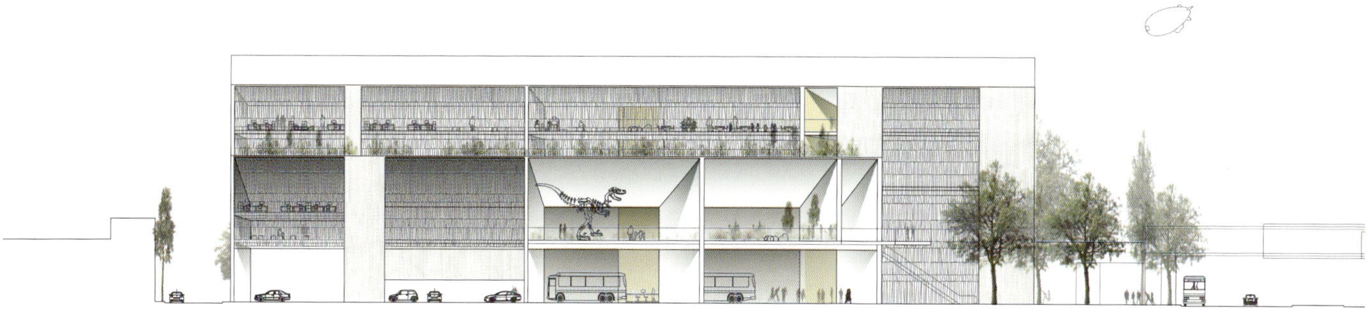

东立面 east elevation

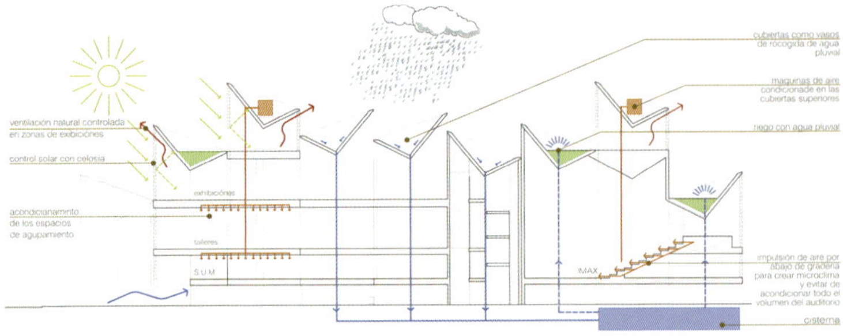

set. The building is set back a few meters from the street to generate an urban plaza, a great public lobby that acts as an extension to the museum. Access to the center is via a diffuse boundary, a forest of column walls that draw the plaza into the inside of the complex.

The public brief of café, gift shop, etc. is located on the ground floor, where there is no access control, and the interactive layout of the exhibition areas located on the upper floors starts on the mezzanine floor, where the multipurpose rooms and foyer of the IMAX cinema and theater are also situated. The different levels of the museum house the various briefs, each with an outdoor space with specific characteristics, such as terraces and gardens.

The building's floor plan is laid out in 10-meter strips that respond to the structural rhythm of the construction module, comprising exposed concrete screen walls that are joined by a V-shaped beam-roof. These walls make up the museum's abstract forest and are designed with careful attention to texture, in a reference to the trunks of the Tree of Tule, with its stout, rugged base that branches out and becomes narrower towards the top. The grouping of modules according to a strategy of horizontal and vertical movement generates the volume of the new museum.

A–A' 剖面图 section A-A'

EXHIBICIONES
TALLERES
CONCESIONES
EXHIBICIONES
GRUPOS
TIENDA

2016年蛇形画廊展馆_BIG

在设计2016年蛇形画廊展馆时，我们试图设计一个包含许多常被人们认为是对立面的东西：形式自由却严谨，模块化但又具雕塑美感，既透明又模糊，可以看作是个盒子，也可以是个无法名状的东西。

我们决定采用砖墙这种最基础的建筑元素，但是砖墙采用的不是黏土砖或石块来建造，而是采用挤压玻璃纤维框架单元层层叠放搭建而成。从底部将墙体拉开，设计成凹洞，供展馆举办展览。墙体像拉链一样被"拉开"，由直线变成曲面，将墙体转换成一个空间，营造出了一个复杂的三维环境，参观者可以里里外外以多种方式进行探索和体验。在最顶部，墙体就像一条直线，而底部形成了有遮挡的山谷，展馆的入口位于谷口，墙面犹如连绵起伏的山坡朝向公园。

"拉开拉链"的墙体形成洞穴般的峡谷，光线通过玻璃纤维框架单元、转体摆放的盒子与盒子之间的缝隙以及玻璃纤维本身半透明的树脂进入展馆。这样，玻璃纤维框架单元的交错叠放和展馆外人们的移动与驻足，都在"洞穴"的墙壁上营造了生动的光与影的演绎。

所使用的建筑材料包括木地板和挤压玻璃纤维板，无论是玻璃纤维交织而成的网状结构还是木纹的起伏线条，都使建筑表面显得温暖透亮，具有线性纹理。

凉亭作为蛇形画廊的重要增建建筑建于画廊的草坪上，将委托国际知名建筑师设计，用建成建筑的形式来展览建筑。蛇形画廊展馆设计和凉亭的设计都要考虑这五座建筑之间的关系，要充分考虑蛇形画廊建筑，也要考虑肯辛顿花园的景观。

Serpentine Gallery Pavilion 2016

For the Serpentine Pavilion 2016, we have attempted to design a structure that embodies multiple aspects that are often perceived as opposites: a structure that is free-form yet rigorous, modular yet sculptural, both transparent and opaque, both box and blob.

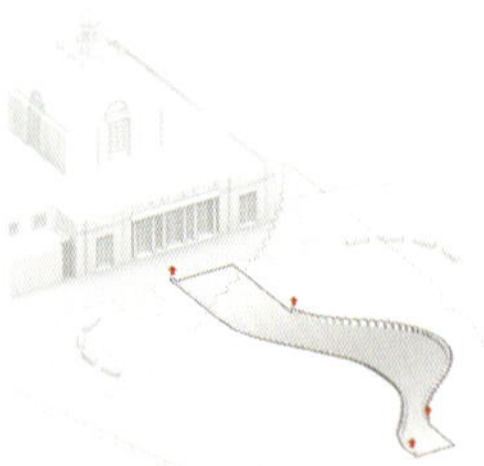

INTEGRATED BENCH
The edges of the path fold upwards to become a continuous bench.

BENCH BECOMES BAR
The bench grows upwards into a bar counter, providing space for the pavilion's cafe.

THE WALL
Exploration of a basic architectural element: the brick wall.

UNZIP
The single wall is pulled apart in a checkered pattern, creating two elevations.

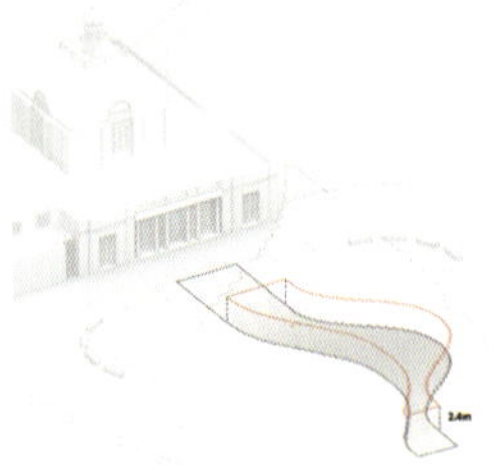

INTERIOR SPACE
Usable space is defined by head-clearance height. The interior pavement is shaped by the curve of the walls, extending on both ends to provide a smooth transition between interior and exterior.

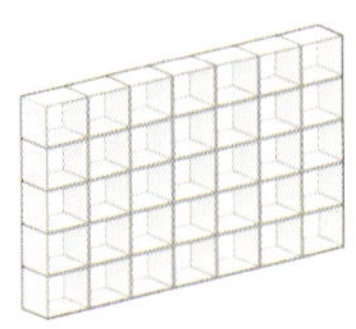

WALL STRUCTURE
Boxes and profiles are arranged in an orthogonal grid.

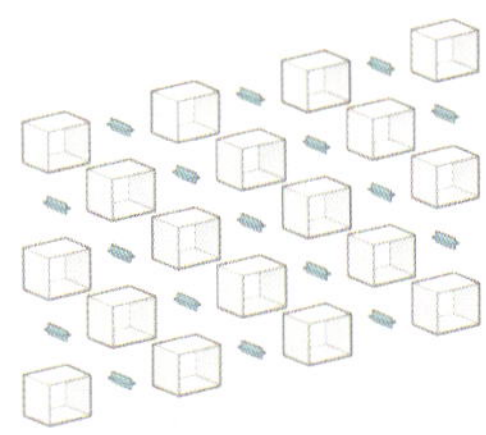

WALL COMPONENTS
The wall consists of 1,802 glass fiber boxes (400mm x 500mm) with 2,890 cruciform aluminum extrusions.

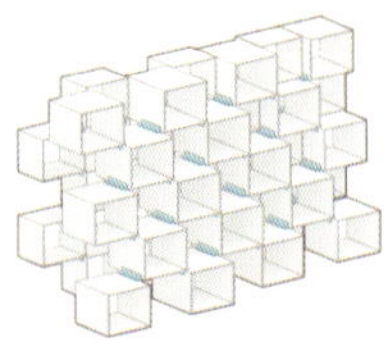

SPATIAL WALL
The boxes slide inwards and outwards in a checkerboard pattern, unfolding in two layers.

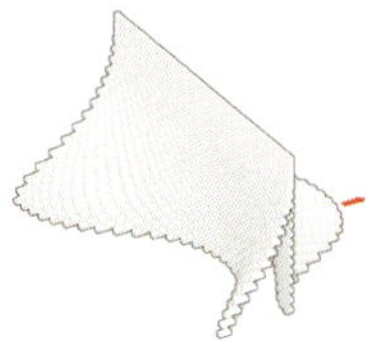

WALL BECOMES SPACE

The straight wall transfroms into two sine curves with an undulating interior.

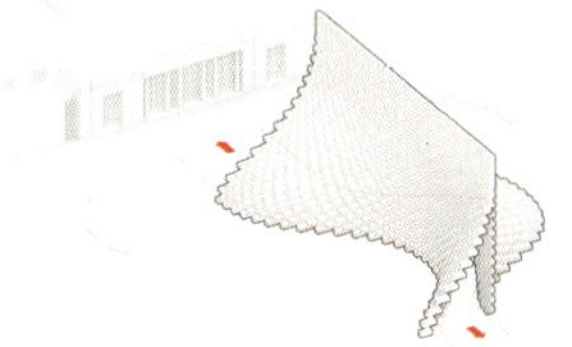

SITE

The pavilion is positioned perpendicular to the gallery. At ground level, the gallery's front terrace and lower walkway are connected via the interior space.

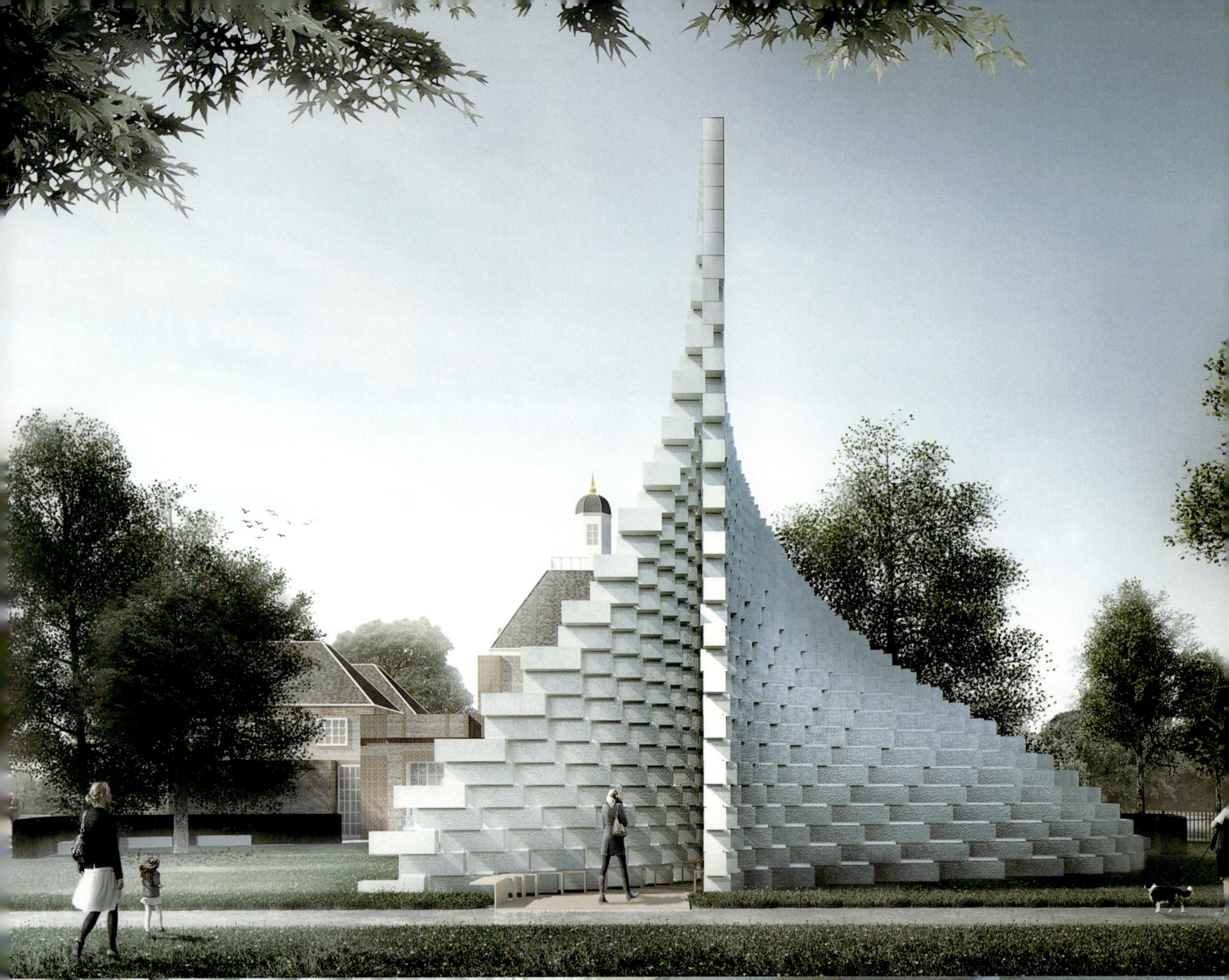

We decided to work with one of the most basic elements of architecture: the brick wall. Rather than clay bricks or stone blocks – the wall is erected from extruded fiberglass frames stacked on top of each other. The wall is pulled apart to form a cavity within it, to house the events of the Pavilion's programme. The unzipping of the wall turns the line into a surface, transforming the wall into a space. A complex three-dimensional environment is created that can be explored and experienced in a variety of ways: inside and outside. At the top, the wall appears like a straight line, while the bottom of it forms a sheltered valley at the entrance of the Pavilion and an undulating hillside towards the park.

The unzipped wall creates a cave-like canyon lit through the fiberglass frames and the gaps between the shifted boxes as well as through the translucent resin of the fiberglass. As a result, the shifting overlaps as well as the movement and presence of people outside create a lively play of light and shadow on the cave walls within.

The materials include wooden floors and extruded fiber line profiles, providing every surface with a warm glow and linear texture – from the mesh of woven glass fibers to the undulating lines of the grain of the wood.

The Summer Houses are a significant addition to the Serpentine's programme of commissioning international architects to design a structure for the Gallery's lawn as an exhibition of architecture in a built form. The designs for the Pavilion and Summer Houses consider the relationships between each of the five structures, the Serpentine Gallery building and the landscape of Kensington Gardens.

绘画工作室：赞美自然光线的工作室

CRAB studio

Bournemouth is about PINES........
P.C.

Most of the built world seems to
be rectilinear........
but not THIS little baby!
P.C.

伯恩茅斯艺术大学副校长斯图尔特·巴塞洛缪以及该校的建筑师校友彼得·库克爵士都十分相信绘画的力量和绘画在艺术创作中的核心作用。2014年，巴塞洛缪先生邀请库克设计了这座专门用于艺术追求的建筑物。当时，库克的事务所CRAB（库克-罗波萨姆建筑事务所）与项目建筑师Jenna Al-Ali共同设计了这个可以供17个院系的学生使用的工作室。这些院系包括绘画、雕塑和插画，还有建筑、服装、戏剧和模型制作等不同学科专业。因此，这个工作室成了这所具有100多年历史的英国艺术院校里第一个拥有独立建筑的专用绘画工作室。

该工作室的构思和设计都是围绕着自然光这一现象进行的，因此它被整体设计成了一个140m²的大房间。这样的话，有一扇30m²的椭圆形朝北的窗户作为主要照明来源，而房间后部的采光来自于10m²的天窗。东侧，位于长凳下方有一排低矮的窗户，进一步将自然光引入室内，散发出神秘的光亮。最后，渐变的自然光透过玻璃大门，照射到远处的弧面墙壁上。

建筑物的外部整齐划一，统一被涂上蓝色，成为伯恩茅斯艺术大学校园北端的"标志性"元素。建筑内部则统一涂上白色，因此可以为绘画活动作衬托。这种一致性是为了让人们更多地关注其建筑形式，以及建筑外部和内部的波动起伏。

这一单体建筑向一侧的一个小小的"门厅"稍稍展开，这个地方对真正的入口起到了保护作用，同时这也激发了设计师的灵感，把形式上与之相关联的另一侧作为紧急出口。小型的室内区域，如洗手间、设备间和绘画储藏室，都位于工作室的南端，都包含在这个富有雕塑感的建筑主体之中。

建筑物的外表是由17块钢板组装而成的薄钢板硬壳式结构：17块钢板先是在工厂进行预制，再运到施工现场，然后用螺栓固定为成套的部件，接着焊接成一整个光滑、防水的外围护结构。建筑物内表面使用了GRG（玻璃纤维增强石膏）技术，直接与建筑外表相呼应，与此同时也营造了一个为学生提供更好服务的、保温隔声的内部空间。

从一开始，CRAB建筑事务所的设计团队（建筑师）、AKT II（结构工程师）、钢板外壳制造商CIG以及总承包商Morgan Sindall就通力合作，讨论在这个狭小的场地上如何进行每一步的运作。

Drawing Studio: A Studio to Celebrate Natural Light

A belief in the power of drawing and its central role in artistic creativity is vociferously shared by Stuart Bartholomew, the Vice Chancellor and Head of Arts University Bournemouth and its architect alumnus, Sir Peter Cook. In 2014 Bartholomew invited Cook to design this building, which is dedicated to that pursuit. Cook's office CRAB (the Cook-Robotham Architecture Bureau) with Jenna Al-Ali as Project Architect have then developed an instrument that is to be used by the students of all of its 17 departments. They range from painting, sculpture and illustration to architecture, costume, drama, and model-making with the result becoming the first such dedicated Drawing Studio as a stand-alone structure to be built in a British art college for more than 100 years.

The Studio celebrates – and is designed around - the phenomenon of natural light. Taking the form of a large room of 140 square meters. Thus it has a 30-square-meter oval north-light window as main illumination, with a "booster" light for

项目名称：Drawing Studio
地点：Poole, Dorset, UK
建筑师：Cook Robotham Architectural Bureau (CRAB studio)
项目团队：Prof. Sir Peter Cook, Gavin Robotham, Jenna Al-Ali
其他顾问团队：AKT II, P3r engineers, PT projects, HED
总承包商：Morgan Sindall
客户：Arts University Bournemouth
用地面积：1,755m²
面积：170m²
竣工时间：2016.3
摄影师：©Richard Bryant (courtesy of the architect)

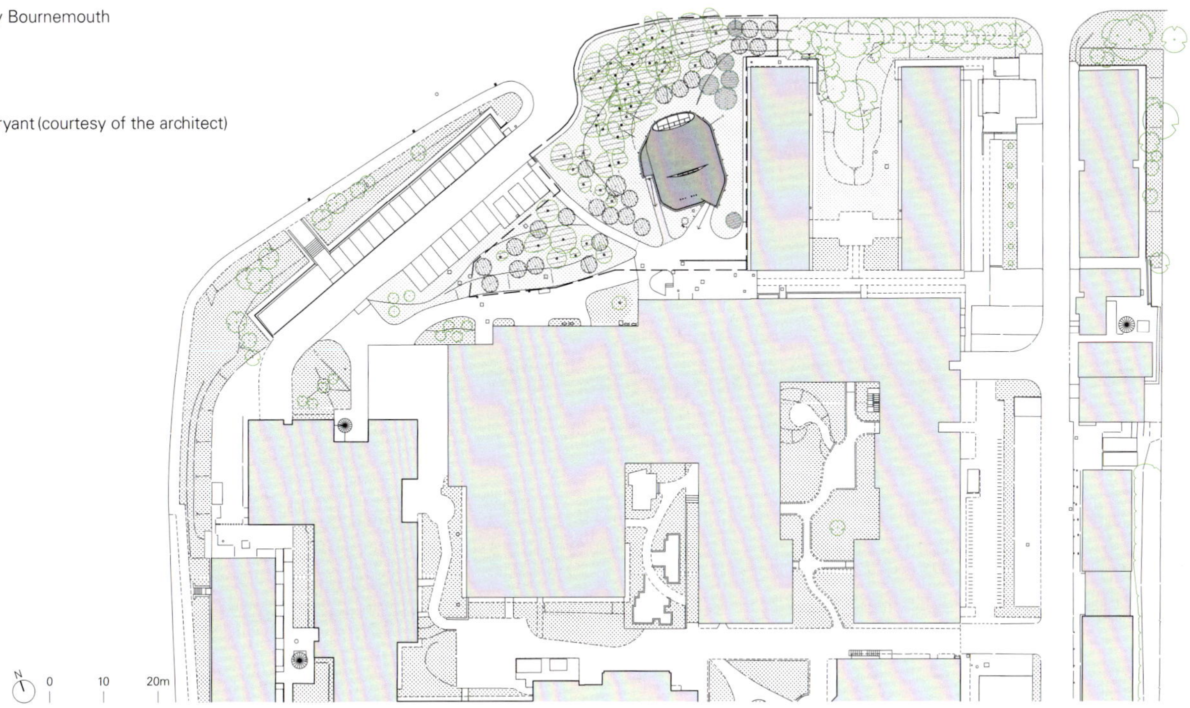

南立面 south elevation

西立面 west elevation

北立面 north elevation

东立面 east elevation

0 5

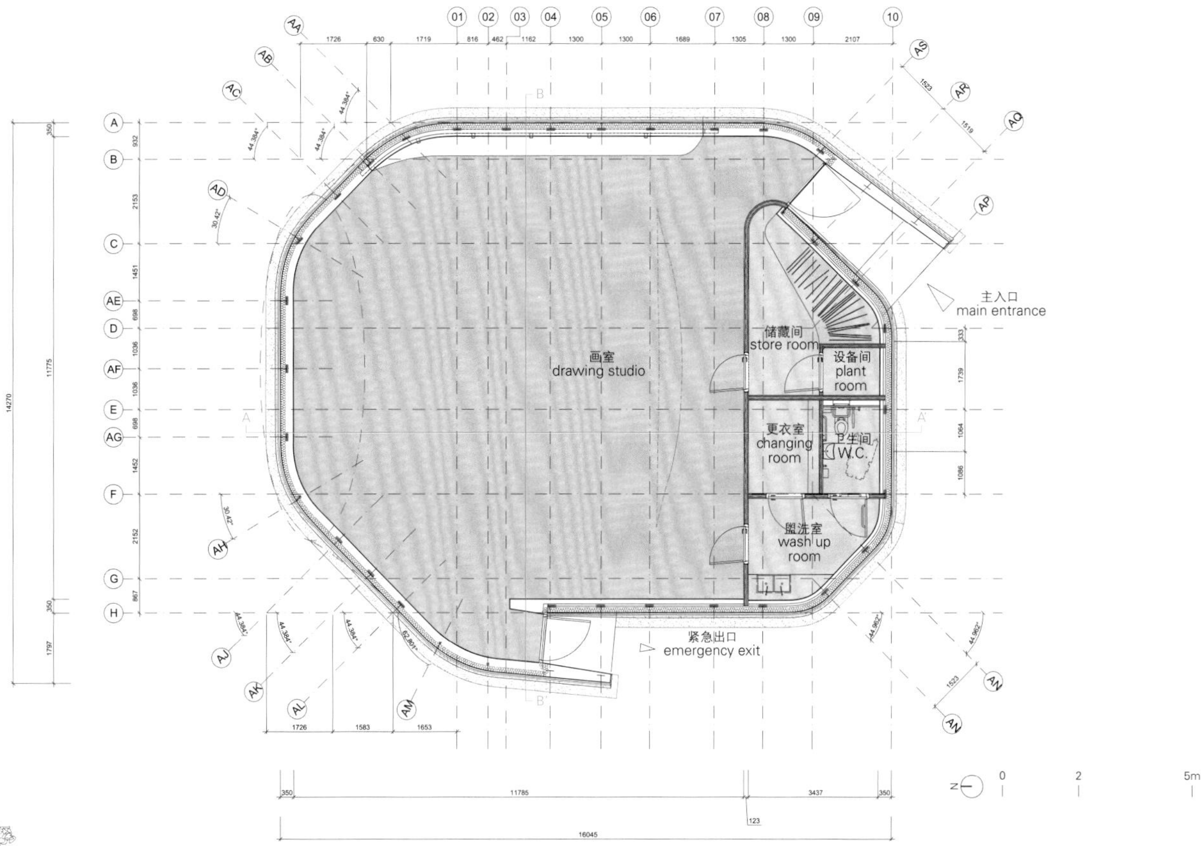

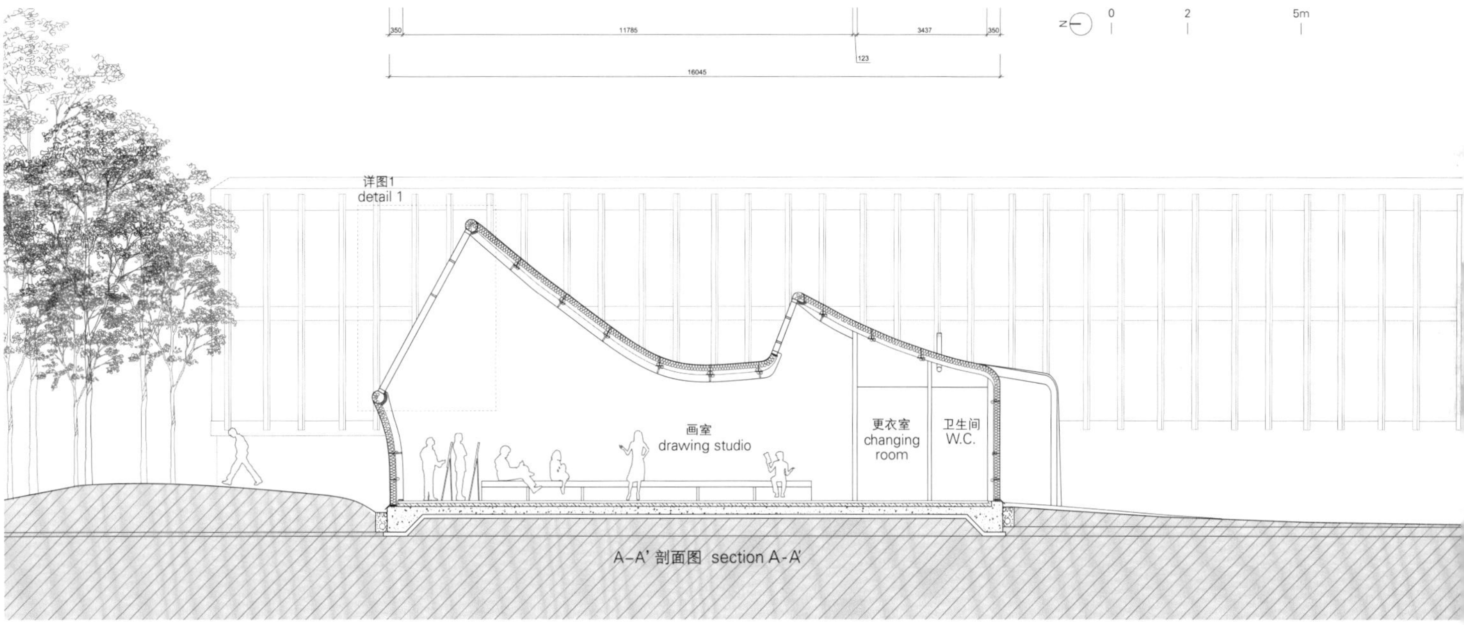

A–A' 剖面图 section A-A'

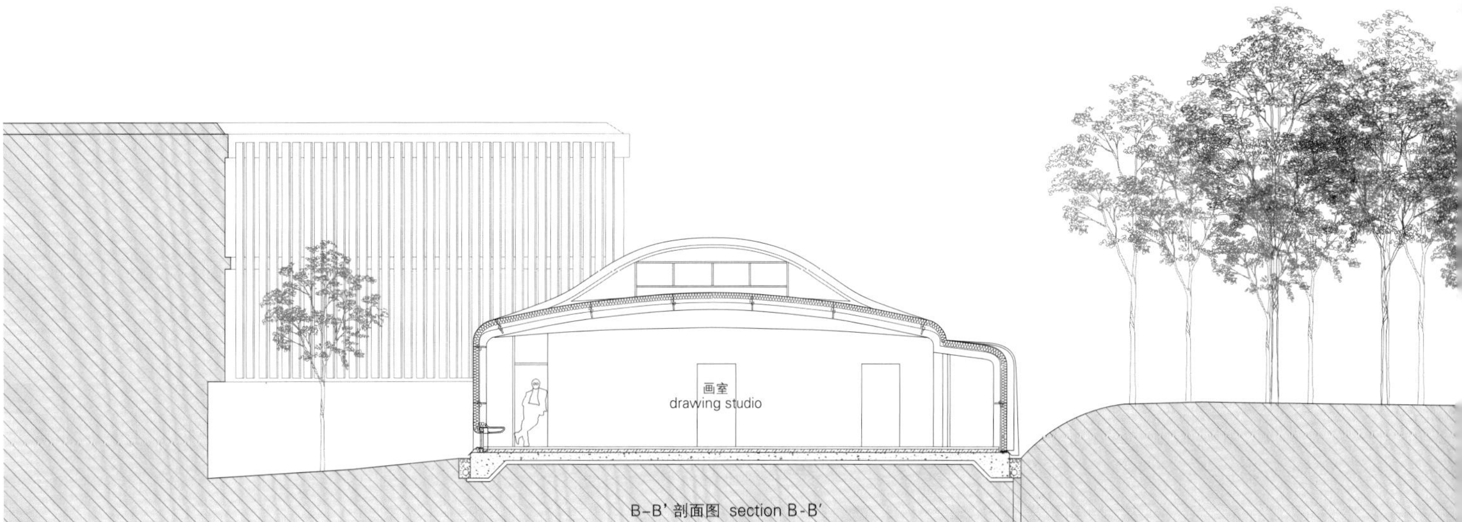

B–B' 剖面图 section B-B'

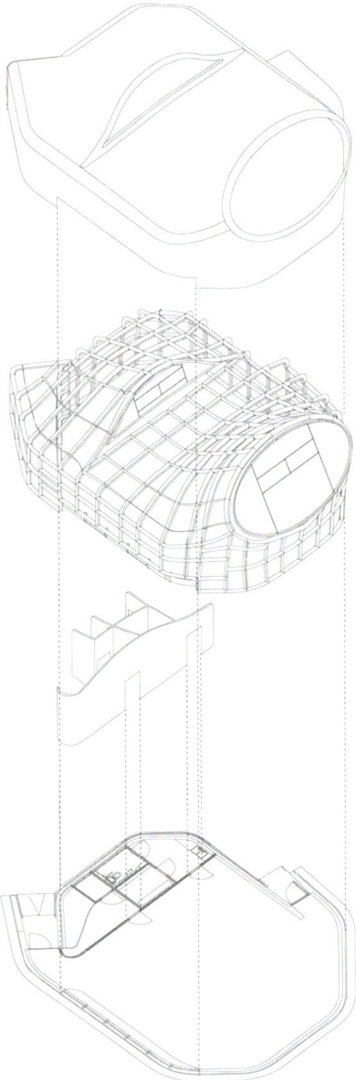

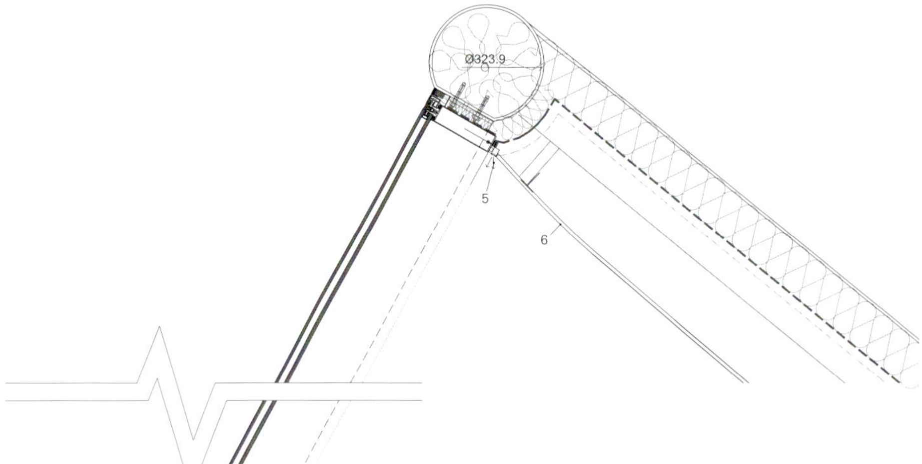

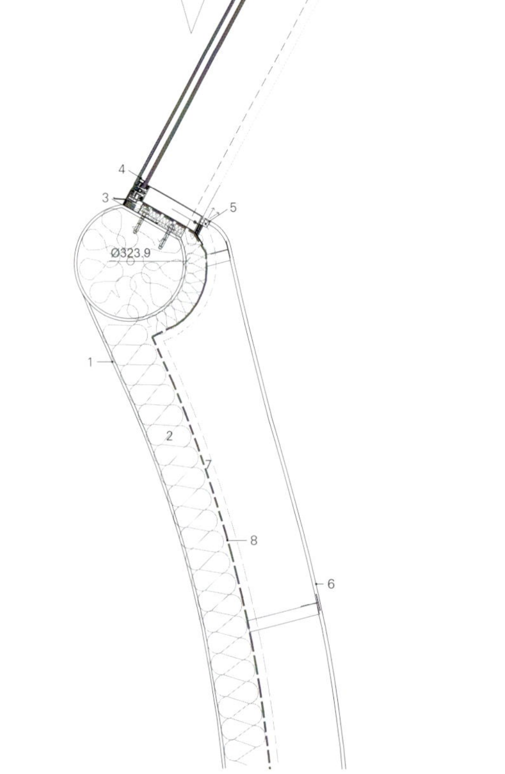

详图1 detail 1

1. 8 mm steel monocoque with blue paint finish
2. 125 mm spray foam insulation
3. deflection head to window fixing
4. structural glazing with vent frame
5. LED flex ught-with 50mm aluminium channel
6. GRG panel-fixing to steel ribs to allow for steel movement
7. 25mm air cavity
8. vapour control layer

the rear of the room from a clerestory window of 10 square meters. On the east side there is a further natural light source in the form of a low window located under a long bench that emanates a mysterious glow and finally, a graduated wash of natural light passes through the glass entry door and along the curved wall beyond.

The outside of the building is a totality – uniformly painted blue and serves as a "marker" element for the north end of the AUB Campus. The interior is uniformly white – and thus acts as a foil to the drawing activity. Such uniformity serves to concentrate the eye upon the form and undulations of both exterior and interior.

This single entity unfolds towards a small "porte clochere" that protects the actual entry door, and inspires a formally related wing for the emergency exit. Small internal zones – the toilets, service room and drawing store room are provided on the southern flank of the Studio room but are encompassed by the major sculptural form.

The external skin is a steel-sheet monocoque construction of 17 steel panels – prefabricated in the factory, transported to site and bolted as a kit of parts and then welded to form a smooth, waterproof enclosure. The internal skin has been

developed through a technique of GRG (glass reinforced gypsum) that enables the exterior profile to be directly echoed within, whilst at the same time creating a useful servicing and insulation cavity.

From a very early stage, the design team of CRAB studio (architects), AKT II (structural engineers), steel shell fabricator CIG and main contractor Morgan Sindall all collaborated very closely and discussed a sequenced operation on a very small site.

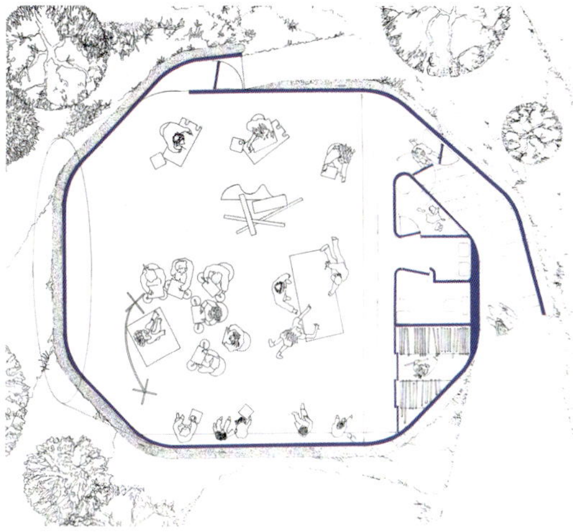

2016年3月，彼得·库克爵士的事务所CRAB（库克—罗波萨姆建筑事务所）完成了为英国伯恩茅斯艺术大学设计140m²的绘画工作室的工作。该工作室将为全校所有的17个院系服务，为来自多个不同学科的学生提供一个独特的空间。这些学科包括绘画、雕塑、插画、建筑、服装、戏剧和模型制作。就该项目的情况，西尔维奥·卡尔塔采访了彼得·库克爵士。

In March 2016 Sir Peter Cook's office CRAB (the Cook-Robotham Architecture Bureau) has completed a 140-square-meter Drawing Studio for the Arts University Bournemouth, in the United Kingdom. The Studio will serve all the seventeen departments of the University, providing a unique space for students across several disciplines, including painting, sculpture, illustration, architecture, costume, drama, and model-making. Silvio Carta interviewed Sir Peter Cook on the project.

能给我们提供一些关于该项目的背景信息吗？

这次设计委托来自于这样一条共识：应该承认绘画的根本性质及其激发思想的方式，需要建造一栋建筑来认可、彰显绘画的价值和作用；同时，这栋建筑可供伯恩茅斯艺术大学所有17个院系使用。这些院系涵盖了从绘画、雕塑和插画到建筑、服装、戏剧和模型制作等不同学科专业。因此，这个工作室将成为这所具有100多年历史的英国艺术院校里第一个拥有独立建筑物的专用绘画工作室。

该工作室的构思和设计都是围绕着自然光这一现象进行的。其外观很简单但却具有鲜明特征，是一座有着优美蓝色曲线的钢结构建筑物。内部，白色的墙面与灰色的地板相协调。通过如此简单的设计，在这个充满活力、忙碌而又具有创新性的艺术院校里，绘画就变成了一种安静、充满想象力的活动。

结构系统和生产工艺似乎是该建筑成功的关键。您能详细地说一下这个问题以及部件预制和组装系统吗？

建筑物的外表皮是由17块钢板组成的硬壳式单体结构。这些部件先在工厂里进行预制，再运到施工现场，然后用螺栓固定为成套的部件，接着焊接成一整个光滑、防水的外围护结构。这种外壳的生产工艺采用了造船业的船体建造方法。因为在造船业中，复杂的曲面外形很常见。这些建造工具的发展使建造环保的建筑外围护结构以及尽善尽美的建筑外形成为可能，建筑外部不再需要单独的覆层，从而节约了人力和物力。

建筑物内表皮使用了GRG（玻璃纤维增强石膏）技术，直接与建筑外表相呼应，与此同时也营造了一个为学生提供更好服务的、保温隔声的内部空间。

Can we have some background information for this project?

This commission came through a recognition that the fundamental nature of drawing and the way in which it stimulated thought needed to be recognised and celebrated in a building available to all 17 sections of the Arts University Bournemouth. They range from painting, sculpture and illustration, to architecture, costume, drama, and model-making with the result becoming the first such dedicated Drawing Studio as a stand-alone structure to be built in a British art college for more than 100 years.

The Studio celebrates – and designed around - the phenomenon of natural light. The exterior is a simple but deliberately identifiable and curvaceous form in blue, constructed from steel. The internal surface is white with a grey floor. Through such simplicity, the act of drawing becomes a calm and considered activity within the lively world of a very busy and creative institution.

The structural system and the production techniques seem to be the key points of this building. Would you elaborate on this as well as the prefabrication and assembly system?

The external skin is a steel-sheet monocoque construction of 17 steel panels - prefabricated in the factory, transported to site and bolted as a kit of parts and then welded to form a smooth, waterproof enclosure. Production techniques drawn from the shipbuilding industry were used to construct the building shell. Complex curved shapes are very common in shipbuilding. The development of these tools has allowed the use of the structure as an environmental envelope and finished architectural form, eliminating the need for a separate external cladding layer and reducing the amount of material and effort.

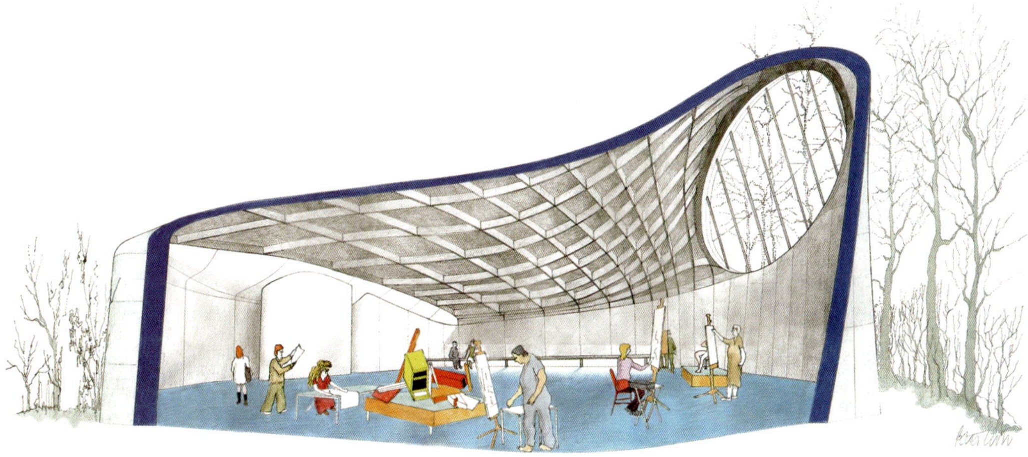

该大学的许多院系都与本项目密切合作，通过这个项目来补充完善对材料之间的空间关系的了解。

该建筑物整体的流线型外观与本地区规整几何形状的建筑形成了鲜明对照，其灵感从何而来？同样，内部洞穴一般的形状与绘画工作室的功能之间有什么关系？

该建筑的外形是由一个能将自然光线引入室内的早期模型演变而来的。我直接想到了艺术家工作室的一贯做法。我们设计了从北面倾斜入射的自然光线，就像你在工厂里见到的光，而不是100年前在艺术家工作室所见到的那种从垂直的大窗户入射的光。但是，该建筑更大，所以我们引入了另外一条来自北部的光线，即设计了一个天窗，可以将光线反射到后墙上。我也想要另外一条光源，从建筑物一侧的长凳下方如涓涓流水溢满室内，以充分利用自然光。

我用巴尔沙木做了一个大模型，人们可以把手放进去推上去或拉下来里面的部件，我记得看到弗兰克·盖里在他的工作室里这样做过。我们把模型放在电脑里，拿给负责建造的AKTII公司的工程师看。整个建筑的鼓起更多是因为结构的原因。

在项目设计的早期，有人就提出这样的想法，与其用基本的木结构加金属板来建，何不全部用金属来建呢？从那以后，设计师就和荷兰的团队一起努力，打造了这样的硬壳式单体结构。整个结构由一系列非常大的钢板组成，钢板的边缘被焊接在一起，就像一个船体，是真正的一体式结构。细节处理都被最小化，没有很多突出来的东西。我最讨厌的就是像英国人那样喜欢用很多不同的建筑材料。我们用的都非常简单，但是有时候这样做起来很难。

The internal skin has been developed through a technique of GRG (glass reinforced gypsum) that enables the exterior profile to be directly echoed within, whilst at the same time creating a useful servicing and insulation cavity.

Many of the departments at the University worked closely with the project using its development to supplement their learning on spatial relationships with materials.

Where does the overall fluid shape in contrast with the regular geometry of the buildings in the area come from? Similarly, is there any relationship between the grotto-like shape of the inside and the programme of this drawing studio?

The building's form essentially emerged by an early model that revolved around the penetration of natural light into the interior. My mind went straight to the tradition of the artist's studio. We established a tilted north light. It's like the light you'd get in a factory, rather than that found in artists' studios 100 years ago, which often had a large vertical window. But this structure was bigger, so we brought in another north light – we had a clerestory that bounces light off the back wall. I also wanted to have another source of light trickling in under the bench which runs along one side. I'm very much playing with natural light.

I made a very large model in balsa so that one could get one's hand inside and push and pull pieces off it very much in a way I remembered seeing Frank Gehry do in his studio. We stuck it on the computer and brought in our engineers, AKT II, and the belly of the profile arched itself more for structural reasons.

Fairly early on, somebody hit on the idea that instead of making this out of a basic timber construction with metal sheeting, why not make it entirely out of metal? And from

Principle of mounting on the building site

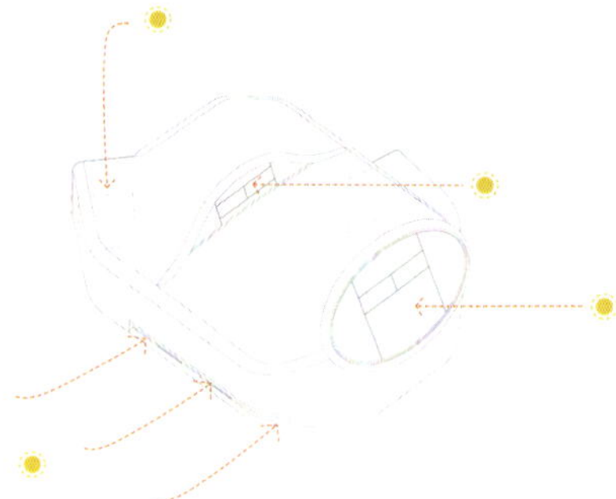

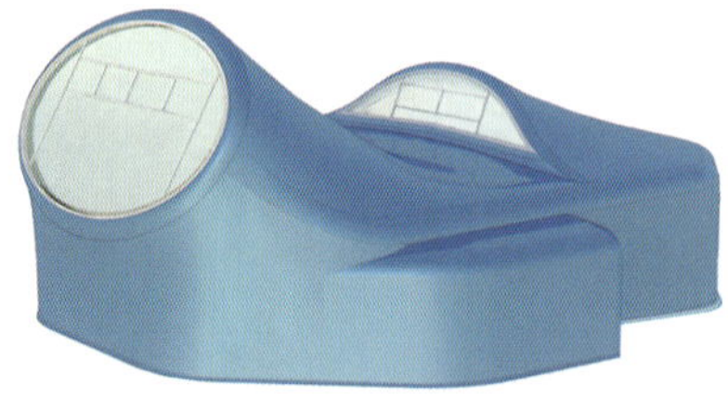

我们对该项目的循环路线及设备间/设施的径向分布特别感兴趣。请给我们介绍一下它的布局。

这是一座独立的建筑，布局非常简单。这一单体建筑向一侧的一个小小的"门厅"稍稍展开，这个门厅对真正的入口起到了保护作用，同时这也激发了设计师的灵感，把外形上与之相互关联的另一侧作为紧急出口。小型的室内区域，如洗手间、设备间和绘画储藏室，都位于工作室的南端，都包含在这个富有雕塑感的建筑主体外形之中。

在该项目中，与自然元素的关系似乎非常重要。例如，自然光的存在为室内带来了独特的氛围。同样，该项目与室外的松树和植被相呼应。您能进一步介绍一下这一点吗？

绘画工作室在设计整体上都结合了可持续发展的理念，与周围起伏的草坡和茂密的树林等自然环境相呼应。围绕在建筑物四周的松树非常重要，所以我们决定保留这儿的松树，在松树之间的空地上安放新建筑。

赞美自然光线的工作室

伯恩茅斯艺术大学绘画工作室的设计看起来是用自然和光线作为推动因素的。适当的光照，包括光的方向、强度、质量，对于一间绘画工作室来说都是非常重要的。光在绘画和雕塑中起着至关重要的作用，不仅影响着艺术家观察其作品的方式，而且影响着观看者的感知方式。在这座硬壳式单体钢结构建筑中，自然光从不同的方向照进室内，营造了一种生机勃勃而又宁静的环境氛围。屋顶上大大的窗户保障了主要的自然光光源，同时，侧墙上的天窗带来了辅助光源，为工作室

there, working with a team out of Holland, came the idea of a monocoque structure. It's a series of very large sheets of steel with flanges that can be welded together. It's put together like a ship, and it's really self-structuring. Detailing is kept to the minimum – there aren't lots of things sticking out. One of my pet hates is the British predilection for using lots of different materials. We have kept it very simple, which is sometimes difficult to do.

We are particularly interested in the circulation and radical distribution of the service spaces/facilities. Please let us hear about the layout.

The building is a free standing pavilion with a very simply layout. This single entity unfolds towards a small "porte clochere" that protects the actual entry door, and inspires a formally related wing for the emergency exit. Small internal zones – the toilets, service room and drawing store room are provided on the southern flank of the Studio room but are encompassed by the major sculptural form.

The relationship with the natural elements seems quite important in this project. The presence of natural light, for instance, confers the interior space with a peculiar atmosphere. Similarly, the project seems to establish a dialogue with the pine trees and vegetation on the outside. Could you expand on this point?

The design for the Drawing Studio incorporates a holistic approach to sustainability. It responds to its natural surrounding consisting of flowing grassed mounds and a dense group of trees. The pine trees that surround the building are very important so we ensure they are remained by locating the building accordingly.

的整体照明发挥了作用。不同光源（不同方向和质量的光）的融合，再加上天花板内弧面的曲线外形，赋予绘画工作室光洁的墙面，艺术家和设计师在此可以专心于创作。墙壁和天花板都没有尖锐的边缘（这一点在剖面上看得尤为清楚），这样光线才能柔和地弥漫开来，不会留下非常明显的阴影区域。较阴暗的区域与最明亮的区域做到无缝过渡。遵循流线型空间的设计理念，内部循环空间的设计可以让学生们沿着大房间的外围移动，中央主要的核心区域可专门用于艺术创作。主入口和出口位于完全相对的两侧，这样是为了突出其圆形循环交通的设计。工作室凭借其流线型外表和无缝设计，在周边的校园中独树一帜，大方地展现它的存在。尽管如此，也可以说是其内部细节的微妙处理，如柔和的自然光和流线型的空间，使工作室成为一个独一无二的空间，里面似乎充满了魔力，激发学生的创造力和想象力。

A Studio to Celebrate Natural Light

The Arts University Bournemouth's Drawing Studio appears to be designed with nature and light as a driving element. The correct amount of light, its direction, intensity and quality are paramount in a drawing studio. Light plays a vital role in drawings and sculptures, influencing not only the way the artist sees his or her work, but also the way the observer perceives it. In this monocoque steel construction the natural light enters the building from different directions creating a vibrant yet calm atmosphere. The large openings on the roof provide the main source of natural light, whilst the clerestory windows on the side walls contribute to the overall illumination of the Studio with a secondary source of light. The combination of different light sources (with different directions and quality of light), together with the curvaceous shapes of the intrados of the ceiling confer this Drawing Studio a smooth finish for artists and designers to concentrate on their creations. The absence of sharp edges in the walls and ceiling (and this is particularly visible in the sections) results in the light being softly diffused, without clearly defined areas of shadows. The darker areas blend transitionally into the brightest seamlessly. Following the idea of a fluid space, the internal circulation has been planned in such way that the students can move around the large room walking along its perimeter, leaving the main central space free for the artworks. The main entrance and the exit are located on diametrically opposed sides to accentuate the circular movement. While this Studio, with its fluid and seamless shapes, establishes its presence clearly and openly in the surrounding campus, it is the subtlety of the details on the inside, such as the softness of natural light and the fluid spatiality, that makes it a unique space with a hint of a magic atmosphere for students to spur their creativity and imagination. Silvio Carta

这座标志性的大厦是由让·努维尔设计的，位于塞浦路斯首都尼科西亚中心，紧邻艾佛雷西亚广场，它改变了尼科西亚的城市轮廓。该项目包括10层的居住公寓、6层的办公空间以及2层的零售区域。从中世纪时期就环绕着该城市的城墙和护城河所具有的水平特征与大厦的垂直特征形成了对比，构成了一系列的反转动作，这也是建筑的特色。这座67m高的建筑将成为该城市最高的地标。

垂直方向上的植物景观覆盖了建筑南立面大约80%的面积。这一特殊的生存环境就如同天然的"遮阳板"。这些植物还扮演着天然的阳光控制系统的角色，夏季可以为公寓和办公室遮挡直射光，冬季又能最大限度地让阳光入射。

这个"活立面"使用了多种多样的塞浦路斯本地攀爬植物，一年四季不断地生长变化。建筑首层周围生长着有400年历史的橄榄树，与垂直立面相互衔接，并与附近的公园互为呼应，成了该城市景观的一部分。

公寓和办公室的特色是都拥有内/外凉廊，充分利用了当地气候温和的特点，将内部生活空间延伸到室外。

在北立面，凉廊延伸至外面，可以最大化地欣赏到公园和城市天际线的美景。一系列连通的露台和大阳台可以让人们在此感受地中海气候，体会徜徉在户外的美好。

东立面和西立面的凉廊从拥有如像素一般图案的混凝土墙向内延伸，以保护居住者和办公室工作人员远离噪声、大风和日晒问题。在混凝土墙上随机开了大量洞口，有的有玻璃，有的没有。尤其在花园区，墙体上有很多洞口，打消了墙体的厚重之感。一天之中，太阳的位置不同，内外空间的光影效果也不一样。这些聚集的孔眼和窗户以0.4m×0.4m为基本尺寸，形成了一幅独特的画面，从内外看都有不一样的景色。晚上，这面墙如同一面巨大的屏幕，显示了居民的用光需求，并进一步加强了这一立面的随机系统。

在大厦最上面的两层，围绕着一个中央庭院设计了一套复式公寓，这是受到了传统的塞浦路斯建筑的启发。三张大遮阳板保护建筑免受夏季炽热阳光的烘烤，同时也凸显了天空的景色。

该项目将会把景观和高层建筑的结合推向一个新的层次，并且为尼科西亚带来一个象征城市未来的新的地标性建筑。

White Walls

The iconic tower designed by Jean Nouvel at the center of Nicosia, next to Eleftheria Square transforms the city silhouette of Nicosia. The program includes 10-story residential apartments, a 6-story office space and a 2-story retail area. This verticality in relation to the horizontality of the medieval walls and the moat that enclose the old part of the city sets the stage for a series of inversions that are characteristic of the building. The 67-meter-high building will be the culminating landmark of the city.

On the south facade a vertical landscape covers approximately 80% of the building's facade area. This exceptional living environment is working like a natural "brise soleil". The plants will act as a natural sun control shielding the apartments and the offices from direct sun during summer while admitting a maximum of sunlight in winter.

This "living facade" supports a variety of Cypriote climbing and spreading plants and will be continually transformed by the cyclic movements of the different seasons. 400 years old

白墙大厦

Ateliers Jean Nouvel

西南立面 south-west elevation

东南立面 south-east elevation

东北立面 north-east elevation

西北立面 north-west elevation

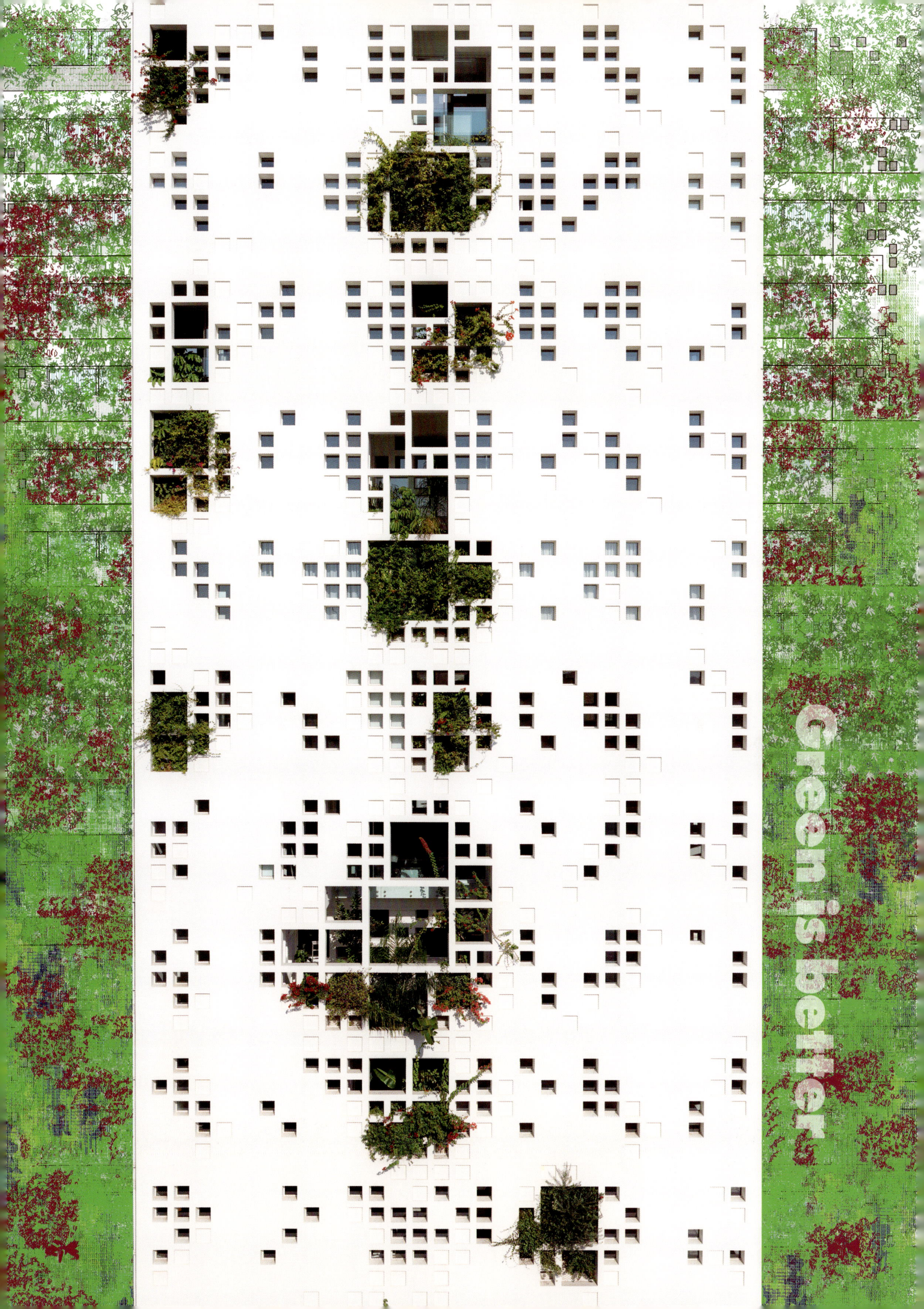
Green is better

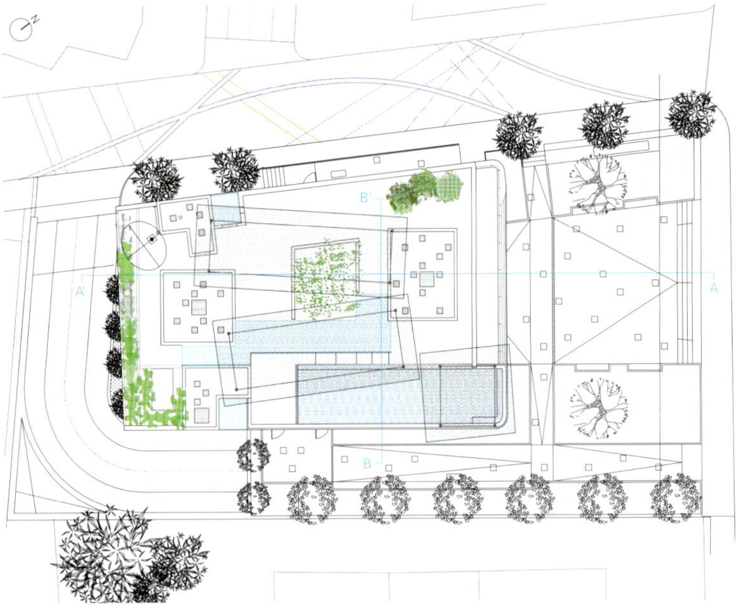

A-A' 剖面图 section A-A'

B-B' 剖面图 section B-B'

项目名称：White Walls / 地点：Stasinou 6, 1065, Nicosia, Cyprus / 建筑师：Ateliers Jean Nouvel / 总建筑师：Jean Nouvel /
项目团队：Elisabeth Kather (Project Leader Studies and Construction), Philippe Papy (Project Leader Concept Design), Sony Devabhaktuni, Anastasia Koumoulidou, Jos Miguel Pomares, Demetris Sophocleous, Nobuo Yoshida / 结构工程师：K. Lissandros_KAL Engineering / 当地合作建筑师：Takis Sophocleous Architects /
顾问：Emmanuel Blamont / 机电水暖消防设计：Mitsides Samouhl & Parntners /
工料测量师：Nicolaou & Konnides / 客户：Nice Day Developments LTD / 用途：retail, apartment and office tower / 用地面积：1,270m^2 / 建筑面积：6,739m^2 /
总建筑面积：10,000m^2 / 建筑楼层：16 floor (6 levels office area, 10 levels apartment area), 67m height / 造价：GBP 25,445,000 (construction cost: GBP 11,500,000) /
施工时间：2007.9—2013.6 / 竣工时间：2015.4 / 摄影师：©Yiorgis Yerolymbos (courtesy of the architect)

olive trees continue this vertical landscape on the perimeter of the building. It creates the link to the adjacent urban park and gets a part of the scenery of the city.

The apartments and office feature indoor/outdoor loggias that extend the living space outside to take maximum advantage of Nicosia's temperate climate.

On the north facade, the loggias extend out to take maximum advantage of views toward the park and the city skyline. A series of cascading terraces and large balconies acknowledge the Mediterranean climate and the significance of spending time outside.

On the east and west facade the loggias extend in from the pixelated concrete walls facade to protect residents and office occupiers from noise, wind and sun. Punctured by a random array of openings, sometimes glazed sometimes left open, especially in the garden areas, the massiveness of the walls is negated by the numerous perforations. Interesting shadow games determine the exterior and interior space in relation to the different sun position of the day. An accumulation of voids and windows shaped by a size of 0.40m x 0.40m creates a unique screen image from the outside and inside point of view. At night the wall as a giant screen is pronouncing the inhabitants light needs and strengthens even more the random system of the facade.

On the top two floors of the tower, a duplex apartment is organized around a central courtyard inspired by the Cypriot traditional architecture. Three large shades protect the apartment from the summer sun and emphasize the view to the sky.

The project will push the integration of landscape and tower architecture to a new level and offer Nicosia a new architectural icon that symbolizes the city's future.

绿色更好
Green is better

"白墙"是塞浦路斯尼科西亚中心的标志性大厦，由让·努维尔建筑事务所（AJN）设计。"Tower 25"大厦高62m，是这个国家第三高的建筑，这是由总部位于美国芝加哥的高层建筑和城市居住区理事会评出的。

该建筑面对着这座首都城市中的威尼斯防御工事，靠近艾佛雷西亚广场，这座广场是由扎哈·哈迪德建筑师事务所重新设计的，目前正在施工。坚实的几何形状的地基将该大厦紧密地融入城市街区的建筑物中，装饰性的"像素"立面减少了它的体量感，变成一种有机的外形。这幢建筑面向西南，每层都有露台，并且采用了一种几何形状，使这个现代化的盒子结构像素化，穿过内部和外部空间之间的界面。这种无规则排列的方形洞口可以说是对于传统塞浦路斯天井概念的当代重塑，即由AJN将其再塑造成凉廊。这些区域为办公区和住宅区共用，和内部花园一起形成了休闲空间。在让·努维尔建筑事务所最近的项目中，由于和帕特里克·布朗（如在澳大利亚悉尼的"中心公园"项目中）等植物学家的合作，绿色模式正在变得越来越重要。在"中心公园"中，这位法国植物学家在立面上嵌入了1200m²的垂直花园，为的是减少太阳得热和平衡"热岛"效应的现象。相反，在尼科西亚的这座大厦中，垂直景观和从像素化白墙上探出头的绿色植物共同构成了天然的太阳光控制系统，大约占到立面的80%。

居住区和办公区的特色是都有露天生活区域，以便在夏天最热的时候提供自然通风。而且16个凉廊也具备这样的功能。建筑顶部（高61.85m）有一个游泳池，在池中可以看到地中海的如画风景。该大厦的

"White Walls" is an iconic tower in the center of Nicosia, Cyprus, designed by Ateliers Jean Nouvel. With 62 meters of height, "Tower 25" is the third tallest high-rise building in the country, stated by the Council on Tall Buildings and Urban Habitat, based in Chicago, USA.

The building is facing the ancient Venetian Fortifications of the capital and sits next to the Eleftheria Square redesigned by Zaha Hadid Architects and currently under construction. A solid geometric base knits the tower tightly into the urban fabric of the city block, while a decorative "pixeling" disposition dematerialises and reduces its volume into an organic form. The building faces southwest with terraces at each floor, adopting a geometric shape that "pixelates" the modernist box to perforate the interface between indoor and outdoor space. This random array of square shaped openings could be a contemporary reformulation of the traditional Cypriot concept of the patio, reformulated by AJN in a veranda. These areas serve both the offices and the apartments, creating informal relationship spaces with the internal gardens.

In the Ateliers Jean Nouvel's latest projects, the green paradigm is becoming increasingly important, as result of the collaborations with botanists such as Patrick Blanc (as in "One Central Park", in Sydney, Australia). In "One Central Park", the Frenchman scientist has inserted 1,200 square meters of vertical gardens in the facade, in order to reduce the peak heat gain and to counterbalance the phenomenon of the "heat island" effect. Instead, in the Nicosia tower, a natural sun-control during the year is provided by a vertical landscape that covers approximately 80% of the facade, and by the greenery, which emerges through the pixelated "White Walls".

Apartments and office units both feature open-air living spaces, to offer natural ventilation during the hottest hours of the

总建筑面积是10000m²，其审美思想效仿了巴塞罗那的一座27层的酒店——"巴塞罗那费拉万丽酒店"，为万豪国际集团所建。

如今，生态、社会和金融领域的全球化趋势推动了绿色建筑思想和实践的发展，成为建筑设计和设计师所看重的核心部分。欧洲类似的建筑，如，由斯丹法诺·博埃里建筑师事务所设计的米兰（意大利）的"垂直森林"或者是由藤本壮介建筑师事务所设计的蒙彼利埃（法国）的"白树"，这些都描绘了类似的"理想环境"城市愿景。西姆·凡·德·雷恩、詹姆斯·瓦恩斯和吕西安·克罗尔很多年前就对这些价值观和实践建立了理论，作为关键人物，他们推动了在特定的建筑领域与环境设计相关的可持续设计。

看到"白墙"，人们可能会想到一系列的"数字化的新艺术派"，即在一百多年后对这一建筑流派进行参数化的复兴。在这座大厦中，有机外形以其有韵律的花朵图案向自然世界表达了敬意，在其中，图案都被塑造成一种植物衍生的形式。因此，大自然在重复和（明显的）随意的概念中得到了赞美。

"白墙"为尼科西亚这座城市提供了崭新、清新的可持续绿色建筑方式。这一建筑涉及了逐渐在全世界崭露头角的一系列垂直花园、"绿色住宅"、城市农场和"游击队"园艺等。

summer days: a function absolved also by the sixteen loggias. At the top (+61.85 m), a swimming pool allows to observe the picturesque landscape of the Mediterranean Sea. The tower features include 10,000 sqm of gross floor area, with an aesthetic concept that resembles the 27-storey Barcelona inn, the "Renaissance Barcelona Fira Hotel" built for the Marriott hospitality company.

Nowadays, global trends of ecological, social and financial sectors are driving forward the philosophy and practice of green buildings as core part for architecture and designers. Similar buildings in Europe, such as the "Bosco Verticale" in Milan (Italy), by Stefano Boeri Architetti or "L'Arbre Blanc" in Montpellier (France), by Sou Fujimoto Architects, illustrate similar ideas for an "Ecotopian" urban vision. These values and practices were theorised many years ago by Sim van der Ryn, James Wines and Lucien Kroll, key-figures that they have promoted sustainable design at the building-specific scale, associated with environmental design.

Looking a the "White Walls" one may think of a sort of a "digital Art Nouveau", like a parametric revival of this architectural style a hundred years later. In this tower, the organic forms pay tribute to the natural world, with its rhythmic floral patterns where the shapes were molded into plant-derived forms. Nature is here celebrated through the notion of repetition and (apparent) randomness.

The White Wall provides Nicosia with a new and fresh approach on sustainability and green presence in the city. This project relates to a series of vertical gardens, "green houses", urban farms and "guerrilla" gardening that are gradually emerging throughout the world. Fabrizio Aimar

二层 first floor

八层 seventh floor

十三层 twelfth floor

六层 fifth floor

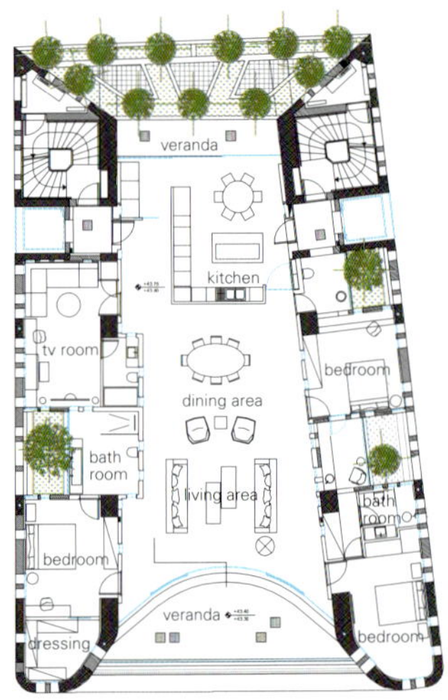

十二层 eleventh floor

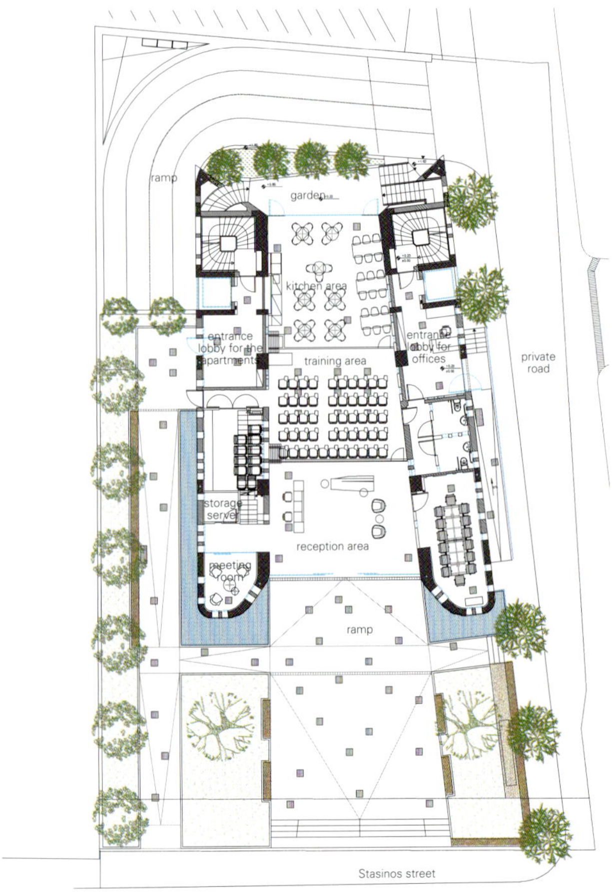

一层 ground floor

四层 third floor

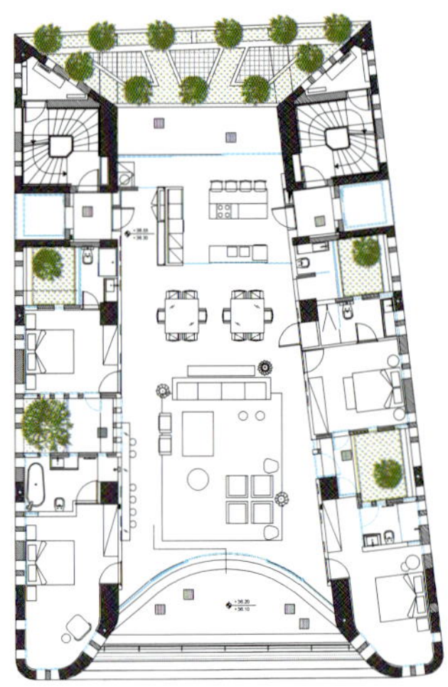

十层 ninth floor

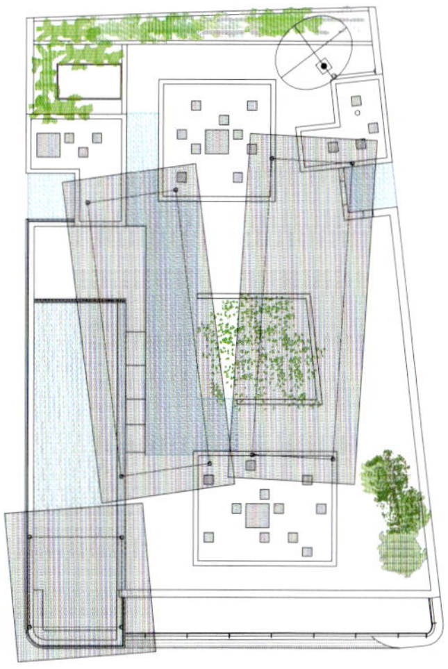

屋顶 roof floor

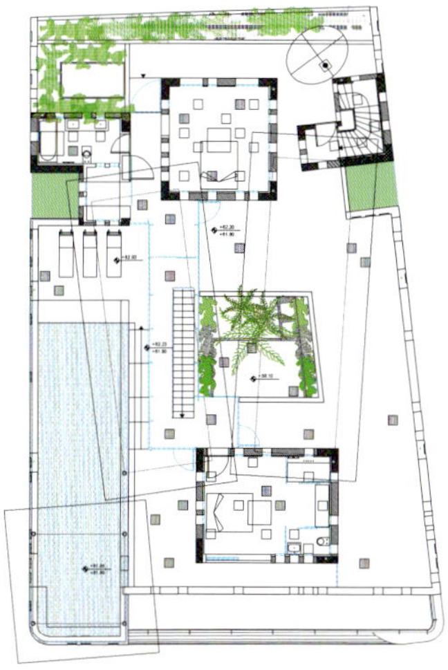

十七层 sixteenth floor

十六层 fifteenth floor

创意工作间
——知识经济时代的办公室设计

The Creativ

Office Design for the Knowledge Economy

在当今的知识经济环境下，创新与知识工作得到了极大的重视，被视为价值创造的驱动力。公司之间相互竞争，招揽顶尖人才，并制定策略培养其主要的创新源泉：个体工作者的创意。如今，公司最关注的就是培养能够打开新视野的这种对话、创造全神贯注的创作环境，并激励员工的参与度，这些也都反映在了工作环境的设计中。通道与过渡空间的设计就是为了促进临时会谈与非正式交谈的发生的，隐私程度不一的工作场所可以让人专心工作，而高品质、可持续发展的工作环境则证实了公司在与其职员共享社会与环境价值观。

Today's knowledge economy places great importance on innovation and knowledge work as the driving force of value production. Companies compete with each other to secure top talent, and devise strategies for cultivating their main source of innovation: the creativity of the individual worker. Fostering the kind of dialogue that leads to new insights, allowing for the concentrated work of the creative process, and motivating worker engagement are primary concerns for companies today, and are reflected in the design of work environments. Circulation areas and in-between spaces are designed to stimulate serendipitous meetings and informal conversation, workplaces with varying degrees of privacy accommodate focused work, and high-quality, sustainable work environments demonstrate that companies share their employees' social and environmental values.

创意工作间——知识经济时代的办公室设计
The Creative Workplace: Office Design for the Knowledge Economy/Isabel Potworowski

e Workplace

创新驱动型经济

工作一直是日常生活的主要组成部分。通过工作，我们展现了自己的创造力，并在圆满完成的工作中得到满足感。然而，工作性质是在不断变化的：20世纪初期，大量的劳动力在工厂工作，而到了现在的社会，逐渐变成以知识工作为主体。

在新兴的知识经济中，知识的生产和开发成为财富与价值创造的重要因素。[1] 在这种经济体系下，是创新驱动了经济增长。新思路、新产品及新技术的发明与实现，让公司获得了新市场，提升他们的竞争优势，因而在很大程度上决定了他们的成功。[2]

首先，组织的创造性取决于个体工作者的创造力。[3] 这要取决于他们理解复杂问题以及从大处着眼、灵活思考提出解决方案的能力。这与好奇心、对不确定事物的承受力以及产生并分享创意的自信心等性格特征都有关系。积极的情感，或是从工作中体会到的快乐，也能通过提高认知灵活性、扩大与给定问题有关的元素的范围来帮助提高创造力。同样，创造力是由内在动机决定的，也是由内在的驱动力、意愿以及对给定任务的投入自主性所决定的。[4]

管理实务的目的就是增强个体积极性，指引工作者的创造力向公司的目标靠拢。提供足够有挑战性的任务、给予工作者选择实现给定目标的方式的自由、组建支持工作组，并提供鼓励与监督机制，这些都是组织管理部门激励员工并激发创造力的方式。[5] 这些努力都体现在了办公环境的设计当中。

办公室如何激发创造力？

办公楼是对工作的意义以及工作如何去执行这些概念的物化。以前，小隔间非常流行，它被当成一种经济的办公室设计方式，可以在一个单独的开放空间内为大批员工提供足够的隐私。然后，为了应对由此

The Innovation-Driven Economy

Work has always been a central part of daily life. Through it, we express our creativity, and we find satisfaction in a job well done. The nature of work, however, is continuously changing: while a large portion of the population worked in factories at the beginning of the last century, today's society is increasingly based on knowledge work.

The emerging knowledge economy is one in which the production and exploitation of knowledge has become the chief factor in the generation of wealth and value.[1] In this economic system, innovation drives economic growth. The invention and implementation of new ideas, products and technologies allow companies to gain new markets, increasing their competitive advantage, thereby largely determining their success.[2]

Organisational inventiveness depends, in the first place, on the creativity of the individual worker.[3] It depends on his or her capacity to understand complex problems and to think broadly and flexibly to propose solutions. It is associated with character traits such as curiosity, tolerance for ambiguity, and confidence in producing and sharing ideas. Positive affect, or the joy that one feels towards his work, also contributes to creativity by increasing cognitive flexibility and the scope of elements that are treated as relevant to a given problem. As well, creativity is determined by intrinsic motivation, and by an internal drive, willingness and freedom to invest oneself in a given task.[4]

Managerial practices are aimed at enhancing individual motivation, steering workers' creativity towards the company's aims. Providing sufficiently challenging assignments, giving workers freedom in choosing the means to achieve a given set of goals, forming supportive work groups, and providing encouragement and supervision are all ways in which organisational management motivates its employees and engages their creativity.[5] These efforts are reflected in the design of the office environment itself.

丹麦Bestseller办公综合楼
Bestseller Office Complex, Demark

造成的单调与孤立之感，又采用了开放式的办公模式，以此来促进员工之间互通有无。然而事实证明，后面这种模式会严重干扰注意力的集中，破坏生产力以及创造性思维。[6]

在当今的知识经济环境下，办公室设计力图打造能更好地培养创造力的环境，并通过创造启发思考、激发创意的空间实现了这一点，为创作过程提供了最佳环境，也通过表达共同价值观来吸引员工。

1. 非正式交流场所

不断有证据表明，不同专业和教育背景的人们相互交流观点和知识能带来全新视角，进而引发有价值的创新。[7,8] 对话交流能拓宽人们的参照标准，还能让人们对既定的任务重燃热情。正是因为有这种动力，古希腊的市集、18世纪英国的茶室以及20世纪初期巴黎的咖啡馆才成了世界上最富有创意的一部分场所。人们聚集在那里交流思想，为道德议题争论不休，还讨论新技术。[9] 简·雅各布斯也描述了曼哈顿西村所具有的这种相似的社会活力：多样化的建筑形式和功能吸引了拥有不同目标的不同人群，而较小的街区距离也产生了各种各样的步行交通。人行道成为人们邂逅、意外碰面交流的场所，由此形成了一种熙熙攘攘的城市环境。[10]

因为公司都希望模仿在城市环境中自发出现的各种各样的非正式交流场景，所以他们的办公室需要具有类似城市的特征。循环交通空间的重要性变得与会议室一样，这是以人行道与附近街道为蓝本的，[11] 而公共广场、公园及咖啡馆也以内部广场、露台及休息室的形式融入了公司的工作环境。

譬如，C.F. Møller建筑师事务所为Bestseller公司在奥尔胡斯设计的新办公综合楼项目中的循环通道，就是围绕位于一层的一条中央"街道"和室内广场而设计的。这是一个三层高的开放空间，人们从主

How Can Offices Engage Our Creativity?

Office buildings are the materialization of notions regarding the meaning of work, and how it should be carried out. Historically, cubicles gained popularity as an economical solution for providing sufficient privacy for a large number of workers in a single open space. Then, as a counter-measure to the resulting monotony and isolation, the open office model was adopted in order to facilitate the flow of ideas and communication between workers. This model, however, proved to be disruptive to attention spans, productivity, and creative thinking.[6]

In today's knowledge economy, office design strives to create environments that are better able to foster creativity. It does so by creating spaces that spark insights and new ideas through interaction and dialogue, provide optimal conditions for the creative process, and engage workers by expressing shared values.

1. Places for Informal Interaction

The cross-fertilization of ideas and knowledge between people of diverse professional and educational backgrounds has repeatedly been shown to generate new insights that lead to valuable innovations.[7,8] Interaction and engaging dialogue can broaden one's frame of reference, and give renewed interest and enthusiasm for a given task. It is through this dynamic that the Ancient Greek Agora, 18th Century British teahouses, and early 20th century Parisian cafes were some of the most creative spaces in the world. People would gather there to exchange ideas, dispute moral issues, and discuss new technologies.[9] Jane Jacobs has described a similar social vitality in Manhattan's West Village: the variety of building types and functions attract diverse people for different purposes, and the short blocks generate a variety of foot traffic. The sidewalk becomes a place of chance meetings and unplanned interactions, generating a bustling urban environment.[10]

As companies strive to emulate the kinds of informal interactions that occur spontaneously in urban environments, their offices acquire city-like characteristics. Circulation spaces gain importance as meeting places modelled on the sidewalk and the neighbourhood street,[11] and public squares, parks and cafes make their way into corporate working environments as internal plazas, terraces, and lounges.

日内瓦JTI总部
JTI Headquarters in Geneva

入口穿过这里可以到达上层的办公室以及餐厅。它位于各条路径的交叉口，所以自然而然地成了一个聚会地点。服装公司也用这里举办时装表演和会议，到那时会有500至1000人聚集在楼梯、上层走廊和"街道"上。在楼内的其他地方，圆形的黑色沙发随意地摆放在主楼梯的附近，为随意交谈和临时会面提供了空间。

Office AT事务所为泰国国家石油勘探生产公司（PTTEP）在泰国北部设计的新办公大楼也非常重视聚会空间。地面层向上折叠，直接借鉴了地下石油勘探的图景，形成了一个坡道景观花园。宽阔的楼梯通往上层的入口，这个中央花园为人们提供了阶梯式的座位，也为会议与活动提供了环境。

SOM建筑设计公司为日本烟草国际集团（JTI）在日内瓦设计的新总部大楼同样也为该公司一千多名员工之间的非正式交流和对话创造了机会。建筑师设计了一条连续的循环路线，依次通过建筑内部的开放式办公室、协同办公区域、咖啡室、健身中心以及许多其他的社交空间。循环路线和会议场所在两个艺术装置的所在地发生了重叠，这两个装置起到了定位点和社交中心的作用。

2. 多样化的工作环境

尽管有创造性的顿悟得益于临时会面和非正式的观点交流，但是创作过程的其他阶段却需要更加私人的环境，以便于专心致志地工作。尤其是回顾、同化研究信息以及孵化（在一个问题的各个方面之间找出潜在的联系，以便于获得可行方案）这样的阶段，都需要精神的高度集中。[12]其他譬如资料收集和方案评估等任务，也需要更多的合作。

此外，如果员工能够随心所欲地选择工作环境，他们会对自己的任务和目标有更强的意识，因此也就会变得更加投入。[13]

Cuantics Creatives Lab在西班牙贝尔梅奥设计的Pevasa总部大楼

The circulation route of C.F. Møller Architects' project for Bestseller's new office complex in Aarhus, for instance, is designed around a central "street" and indoor plaza on the first floor. It is a three storey-high open space across which people pass to reach the upper-level offices as well as the restaurant from the main entrance. Its location at the intersection of different paths makes it a natural meeting place. It is also used by the clothing company for fashion shows and meetings, during which between 500 and 1,000 people can gather on the stairs, upper galleries and on the street level itself. Elsewhere in the complex, circular black sofas are informally arranged around the main staircases, accommodating casual conversations and unexpected meetings.

Office AT's design for Thai national petroleum company PTTEP's new office building in northern Thailand also places importance on gathering spaces. The ground plane is folded upwards, at once making reference to underground oil exploration, and forming a sloping landscaped courtyard. With its broad stairs leading to the upper level entrance, this central garden offers stepped sitting places and a setting for meetings and events.

SOM's design for JTI's new headquarters in Geneva likewise creates opportunities for informal interaction and dialogue among the company's 1,000+ employees with a continuous circulation route that loops through the building, connecting open offices, collaborative work areas, coffee points, a fitness center, and a number of other social spaces. Circulation paths and meeting places also overlap at two curated art installations that serve as points of orientation as well as social hubs.

2. A Variety of Working Environments

While creative insight benefits from serendipitous meetings and informal exchanges of ideas, other phases of the creative process require more private environments for concentrated work. In particular, phases such as reviewing and assimilating researched information, as well as incubation – making subconscious connections between the various aspects of a problem to arrive at possible solutions – require a high degree of concentration.[12] Other tasks such as data collection and the evaluation of solutions entail more collaborative work. Moreover, when workers are free to choose their working environment, they become more conscious of their task and

1. Ian Brinkley, "Defining the knowledge economy" (London: The work foundation, 2006), 3-4, http://www.theworkfoundation.com/assets/docs/publications/65_defining%20knowledge%20economy.pdf
2. James Moultrie, Mikael Nilsson, Marcel Dissel, Udo-Ernst Haner, Sebastiaan Janssen and Remko Van der Lugt, "Innovation Spaces: Towards a Framework for Understanding the Role of the Physical Environment in Innovation," Creativity and Innovation Management 16 (2007): 53, doi: 10.1111/j.1467-8691.2007.00419.x.
3. Wikipedia, "Creativity," https://en.wikipedia.org/wiki/Creativity (accessed April 3, 2016).
4. 5. Ibid.
6. Maria Konnikova, "The Open-Office Trap," The New Yorker, January 7, 2014. http://www.newyorker.com/business/currency/the-open-office-trap

也在开放式和私人工作空间之间取得了平衡。其正方形的开放式平面被五个圆形空间所打断，这几个空间均由圆柱形构件构成。其中一个是花园，有两个是封闭的会议室，提供了更加密闭的工作环境。上层有一个较大的会议室，朝屋顶花园开放，是为公司聚会准备的。

3. 价值重心

员工能否心甘情愿地投入工作以及在工作中能获得多大的快乐，也要取决于组织价值与个人价值相合的程度。例如，20世纪50年代，人们对于工业发展、技术进步与经济增长抱有一种普遍的信念。这些价值观体现在时髦的钢材与玻璃大楼中，在这些大楼中以桃花心木装修的会议室和安装了空调系统的办公室比比皆是。这些大楼展现了高效、管理严谨与极致完美的形象。[14]如今，新一代知识工作者正在寻求一种注重身心和谐的生活方式，重视快乐甚于钱财。对他们而言，一家公司的目标和社会价值与经济增长水平一样重要。[15]人性化尺度、社交联系以及对环境与员工福祉的关心，都成为重要的组织机构特征，也都反映在了办公室设计中。

由RAU建筑师事务所设计的Alliander公司位于荷兰德伊芬的新总部大楼就证实了办公楼如何能体现以上这些价值。Alliander作为一家能源电网公司，将其总部打造成为一个节能综合体。它重新利用场地原有的六座大楼，将它们嵌入装有波浪形屋顶的大型中庭的玻璃墙内。有保温效果的第二层表皮被添加于原有的建筑上，利用的是从现场找到的老旧的电缆盘、电线杆上回收的木材。同样，停车场上方的太阳能板收集了充足的能源，再重新分配一部分给当地电网。此外，中庭内的自然通风系统、充足的自然光线以及绿植，都体现了公司对环境、对员工福祉的关心。

Estudio Carme Pinós在墨西哥设计的Cube 2大楼，展示了一个全

its purpose, in this way becoming more engaged in it.[13]

Cuantics Creatives Lab's Pevasa Headquarters in Bermeo, Spain also provides a balance of open and private workspace. Its square open plan is interrupted by five circular spaces, defined by the building's cylindrical structural elements. One of these spaces is a garden, and two are enclosed meeting rooms that provide a more enclosed working environment. The upper level has a larger meeting room for company gatherings that opens onto a roof garden.

3. Value Emphasis

The willingness to invest oneself, and the amount of joy that one has towards work also depend on the degree of correspondence between organisational and personal values. In the 1950s, for instance, there was a widespread faith in the progress of industry, technology and economical growth. These values found expression in sleek steel and glass towers with mahogany meeting rooms and air-conditioned offices. These towers embodied the image of efficiency, control, and ultimate perfection.[14] Today, a new generation of knowledge workers is seeking a holistic lifestyle, and values happiness over money. For them, a company's purpose and social values are as important as economical growth.[15] Human scale, social contact, and care for the environment and employee well-being are becoming valued organisational characteristics, and ones that are reflected in office design.

Alliander's new headquarters in Duiven, the Netherlands, designed by RAU Architects, exemplifies how an office building can embody such values. As an energy grid company, Alliander has made its headquarters into an energy-positive complex. It re-uses the site's six existing buildings, embedding them in the glass walls of a large atrium with an undulating roof. An insulating second skin is added to the original buildings, using reclaimed wood from old cable coils and utility poles found on the site. As well, the solar panels over the parking spaces gather sufficient energy to redistribute a portion to the local grid. Moreover, the system of natural ventilation, the abundant natural light and vegetation in the atrium expresses concern for the environment and for employees' well-being.

Cube 2 by Estudio Carme Pinós in Mexico, exemplifies a new model for commercial office tower. This project is naturally

7. Kerstin Sailer, "Creativity as social and spatial process," Facilities 29 (2011): 8, doi: 10.1108/02632771111101296.
8. Jonathan Molloy, "Can Architecture Make Us More Creative? Part II: Work Environments," May 10, 2013, ArchDaily, http://www.archdaily.com/367700/can-architecture-make-us-more-creative-part-ii-work-environments
9. Jonathan Molloy, "Can Architecture Make Us More Creative?" April 3, 2013, ArchDaily, http://www.archdaily.com/353496/can-architecture-make-us-more-creative/
10. Jonathan Molloy, "Can Architecture Make Us More Creative? Part II: Work Environments"
11. Ibid.
12. Graham Wallace, cited in Kirsten Sailer, 9.
13. Belinda Lanks, "Cozy in Your Cubicle? An Office Design Alternative May Improve Efficiency," Bloomberg, September 19, 2014, http://www.bloomberg.com/news/articles/2014-09-18/activity-based-working-office-design-for-better-efficiency
14. Juriaan van Meel and Paul Vos, "Funky offices: Reflections on office design in the 'new economy'," Journal of Corporate Real Estate 3 (2001): 327-328, doi: 10.1108/14630010110811661.
15. CBRE Workplace Strategy team, "Fast Forward 2030: The Future of Work and the Workplace," Genesis Research Report (2014): 2, http://www.cbre.com/research-and-reports/future-of-work
16. Raymund Ryan, "Cube II Tower in Guadalajara, Mexico by Estudio Carme Pinos," Architectural Review, June 11, 2015, http://www.architectural-review.com/today/cube-ii-tower-in-guadalajara-mexico-by-estudio-carme-pins/8682736.

新的商业办公大楼模型。该项目有自然通风，并为办公室员工提供了舒适宜人的屋顶露台。还在街面上建了好多花园，为城市带来了绿色空间，与附近其他的以大门和保安小屋与街道隔离开的高楼大厦形成了对比。[16]

Jacques Ferrier建筑设计公司在上海设计的仪电集团办公综合楼也通过一个由六座办公楼组成的"城中村"展示了社会与环境价值。办公楼坐落于一个起伏的景观台地上，下方是商业与公共空间。浑圆的拐角、不规则的排布以及较低的楼高，都赋予综合体友好、人性化的姿态，跟住宅区差不多。同样，该项目也用栽种绿植的路堤和绿树成荫的公共通道替换了附近河道边的灰色围墙，把自然环境还给了当地社区。

由Piuarch设计、位于米兰的杜嘉班纳公司办公楼则通过材料与图案的运用，展现了这家时装公司的独特风格与审美价值。面向皮亚韦街一侧立面上的方形不锈钢百叶模块的重重叠叠、令人惊叹的花纹，与灰色玻璃面板的彩色变化一起，营造了一种活泼生动而又奢华的表面。地面铺的是玄武岩，这是一种黑色的火山岩，体现了杜嘉班纳公司的地中海特色。

其办公室设计鼓励临时会面和偶然交谈，提供了隐私程度各有不同的多种工作环境，而且通过表达与新生代知识工作者产生共鸣的价值观，促使员工发挥创造力，激励他们更加投入地工作，为实现公司的集体目标而努力。尽管创造力本身是源自于个人的，不可能由管理架构或空间结构产生，但是这些外部因素确实在激发和增强个人创造力方面起到了重要的作用。因此，一间精心设计的办公室能引导员工的创造力发挥在具有社会价值和经济价值的创新制造上，而且同样重要的是，让我们得以在日常工作中表现自我、体会到满足感。

ventilated, and offers pleasant roof terraces for office workers. It has planted gardens at street level, and offers green space to the city, in contrast to the other high-rises in the neighbourhood that are walled off from the street with gates and security cabins.[16]

The Yidian office complex in Shanghai by Jacques Ferrier Architecture also demonstrates social and environmental values with a "village" of six office buildings that sit on an undulating landscaped plateau with commercial and common functions below. The rounded corners, irregular arrangement and low building height give the complex a friendly, human scale with an almost residential character. As well, the project replaces the grey walls along a nearby canal with a planted embankment and tree-lined public walkway, giving the natural environment back to the local community.

Dolce & Gabbana's office in Milan, designed by Piuarch, expresses the fashion house's stylistic and aesthetic values by its use of materials and patterns. The overlapping, staggering pattern of square stainless steel brise-soleil modules on the Viale Piave facade, together with the chromatic variation of grey glass panels, creates a surface that is at once animated and luxurious. The ground floor is laid with basaltina, a dark volcanic stone, which expresses Dolce & Gabbana's Mediterranean character.

By encouraging serendipitous meetings and dialogue, providing a wide range of working environments with varying degrees of privacy, and by expressing values that resonate with the new generation of knowledge workers, office design can engage workers' creativity, and motivate them to invest themselves in an organisation's collective aim. Although creativity itself originates from the individual person and cannot be generated by managerial or spatial structures, these external factors do play an important role in engaging and amplifying individual creativity. A well-designed office, therefore, can channel workers' creativity towards the production of socially and economically valuable innovations and, perhaps just as importantly, allows us to express ourselves and to find satisfaction in our daily work. Isabel Potworowski

日本烟草国际集团总部

Skidmore, Owings & Merrill LLP

日本烟草国际集团（JTI）的新总部位于日内瓦地区（该地区有许多久负盛名的国际组织），建在一个以前的工业用地之上，周围绿地环绕。在公司建筑设计团队、结构工程团队和室内设计团队的通力协作下，该竞赛获奖方案将JTI原来的四栋建筑物整合成一座地标性建筑，充分体现了SOM建筑设计事务所的恪守承诺的做法，即致力于综合设计、可持续发展和创新的工作场所设计方案。

该建筑风格不仅考虑到其周边环境，而且考虑到了日内瓦湖和阿尔卑斯山，既敏感地注意到了周边低矮的环境，又在周围赫赫有名的建筑中建立了一种强烈的存在感。给设计带来挑战的三角形地块和最大化提升可用空间的需要决定了建筑物的外形。通过抬升建筑东北角和南角，公共庭院得以形成，有了贯穿场地的通透性，不仅增强了建筑对周边的开放性，还提升了行人直接通往当地交通枢纽的便捷性。令人印象深刻的悬臂式空间和分别达到60m和75m的净跨，构成建筑物独一无二的立面，也使公司员工和参观者到来时有一种独特的感觉。

建筑内的循环路线呈连续的环形——从入口大厅到全景饭店——连接着开放式平面布局的办公室、66间会议室、31个协同办公区域、23个咖啡站、熟食店、商务中心和礼堂、健身中心、2个屋顶露台及许多社交空间。这条环路的独特之处在于色彩的一致性，使用当地材料，表面光滑坚固。为了更好地体现环路的连通性，特邀国际著名艺术家利亚姆·吉利克和索尔·勒维特设计了一系列有教育意义的艺术装置，同时也起到了社交中心和指路标志的作用。

室内设计理念是将水平与垂直的相互连通性最大化，实现项目委托人为其1000多名员工创造一个社交和协作社区空间的需求。SOM建筑设计事务所设计了为整个JTI总部空间量身定制的工作场所模型，打破该公司原有建筑的筒状布局，创造了一个无缝连接的工作场所，为适应JTI以后的长远发展需求提供了灵活性。

通过设计一个外围扭力管钢结构系统这一创新的工程设计解决方案，建筑师建造了跨度为18m的无柱楼面，使开放式平面布局的空间得以实现。

该设计采用了与德国高德纳工程咨询公司合作开发的极具创新性的封闭式腔体立面（CCF），既是一个组合型的幕墙系统，能够根据季节变化对不同的外部气候条件做出响应，也能让人欣赏到室外独特的风景，使阳光最大限度地射入工作场所。通高的玻璃幕墙由一个三层玻璃的内层和一个单层玻璃的外层组成，每块玻璃都3m宽、4.2m高。在内层玻璃和外层玻璃之间形成了一个空腔，中间安装了一个织物面料的卷帘。玻璃面板全部密封，其内部填充了经过过滤和除湿的空气，可以防止空腔内空气的热胀冷缩。CCF系统将使用者的舒适度摆在了第一位，并致力于降低整个建筑的能源消耗和碳排放，促使建筑达到欧洲能源指标和瑞士Minergie可持续性等级要求。这个系统代表了SOM公司历史上最先进、最优秀的全玻璃幕墙系统。

由于对社区产生了积极的作用和影响，这一建筑被人们广泛认可，同时人们希望这一令人梦寐以求的设计将会成为该地区未来发展的催化剂。该项目可以称得上是欧洲最具有可持续性的建筑之一，还包括容纳104名儿童的1162m²的日托中心以及与日内瓦市共同修建的面积达8162m²的新公共空间。

l'Ariana Park
United Nations
Botanical Garden
N
Square of Nations
Park Barton
Park Moynier
Lake Geneva
Park de Vermont
Park Mon Repos

JTI Headquarters

Surrounded by parkland and built on a former industrial site, the new JTI Headquarters is located in a Geneva district home to prestigious international organisations. A collaboration between the firm's architecture, structural engineering and interior teams; the competition-winning design consolidates four existing JTI premises within a single landmark building and demonstrates SOM's commitment to integrated design, sustainability and innovative workplace solutions.

The architecture draws references not only from its immediate context, but also Lake Geneva and the Alps, establishing a strong identity amongst its illustrious neighbours while responding sensitively to its low-rise context. The building's form was dictated by the shape of the challenging triangular site and the necessity to maximise useable space. By elevating the north eastern and southern corners of the building, a public courtyard is carved out, creating permeability through the site, opening it up to the local neighbourhood and facilitating direct pedestrian connections to a local transport hub. The impressive cantilevered space and clear-spans of up to 60m and 75m respectively, provide the building with its unique elevations and also create a unique sense of arrival for employees and visitors.

A continuous circulation route loops through the building - from the entrance lobby to the panoramic restaurant - connecting open-plan offices, 66 meeting rooms, 31 collaborative work areas, 23 coffee points, deli, business center and

项目名称：JTI Headquarters / 地点：Geneva, Switzerland / 建筑师：Skidmore, Owings & Merrill / 项目经理：Knight Frank
室内设计师：Skidmore, Owings & Merrill / 立面工程师/设计师：Skidmore, Owings & Merrill / 本地建筑师：Burckhardt+Partner AG
记录工程师/本地工程师：Ingenie SA / 立面承包商：Josef Gartner AG, Arlesheim / 设备工程：MAB-ingénierie SA, Morges/Weinmann-Energies SA, Echallens / 土木工程：B+S Ingénieurs Conseil SA, Genève / 承包管理：CBRE / 承包商：Implenia Suisse SA / 总建筑面积：28,000m^2 (9 floors with varying floor areas) / 一层建筑面积：2,500m^2 / 高度：51m (apex) / 停车场：158 vehicle spaces, 150 bike spaces,150 motorcycle spaces / 立面：1,277 glass panels, 143 different sizes / 标准面板：1.5 tonnes and 3 x 4.20 meters high / 施工时间：2012—2015
摄影师：©Adrien Barakat (courtesy of the architect) - p.60~61, p.62, p.63, p.67, p.68~69, ©Hufton+Crow (courtesy of the architect) - p.66, p.70, p.72, p.73, p.74, p.75, p.77, ©Johannes Marburg (courtesy of the architect) - p.64, p.65

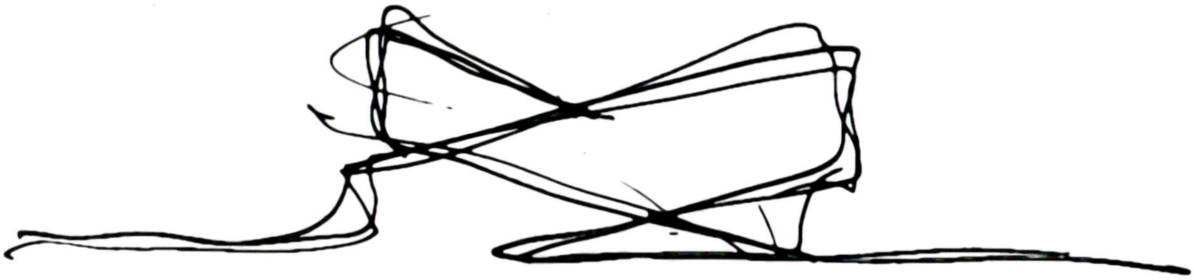

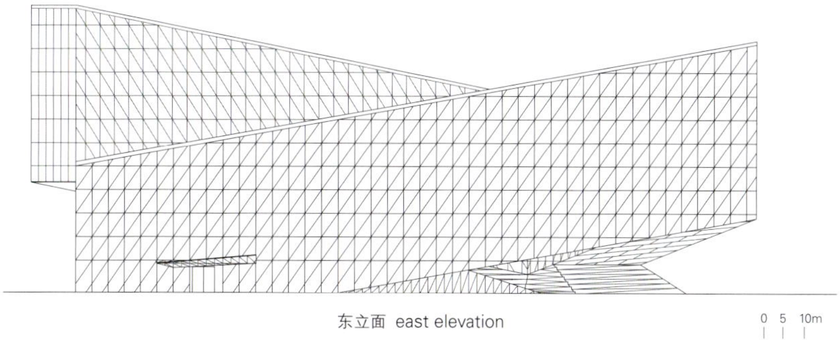

东立面 east elevation

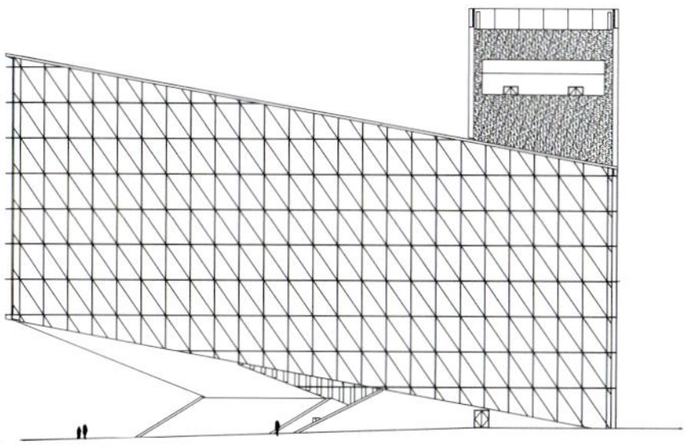

北立面 north elevation

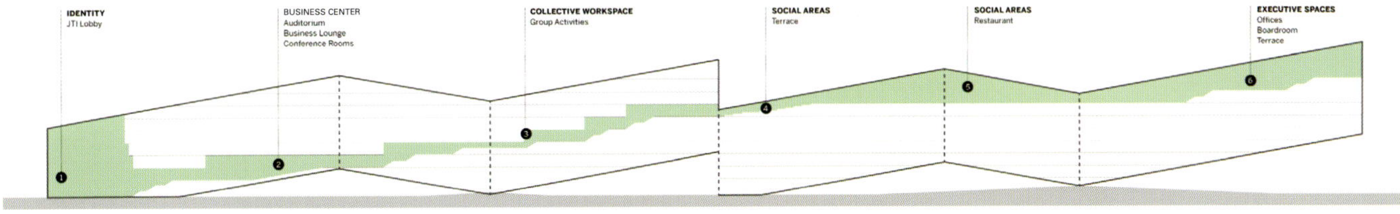

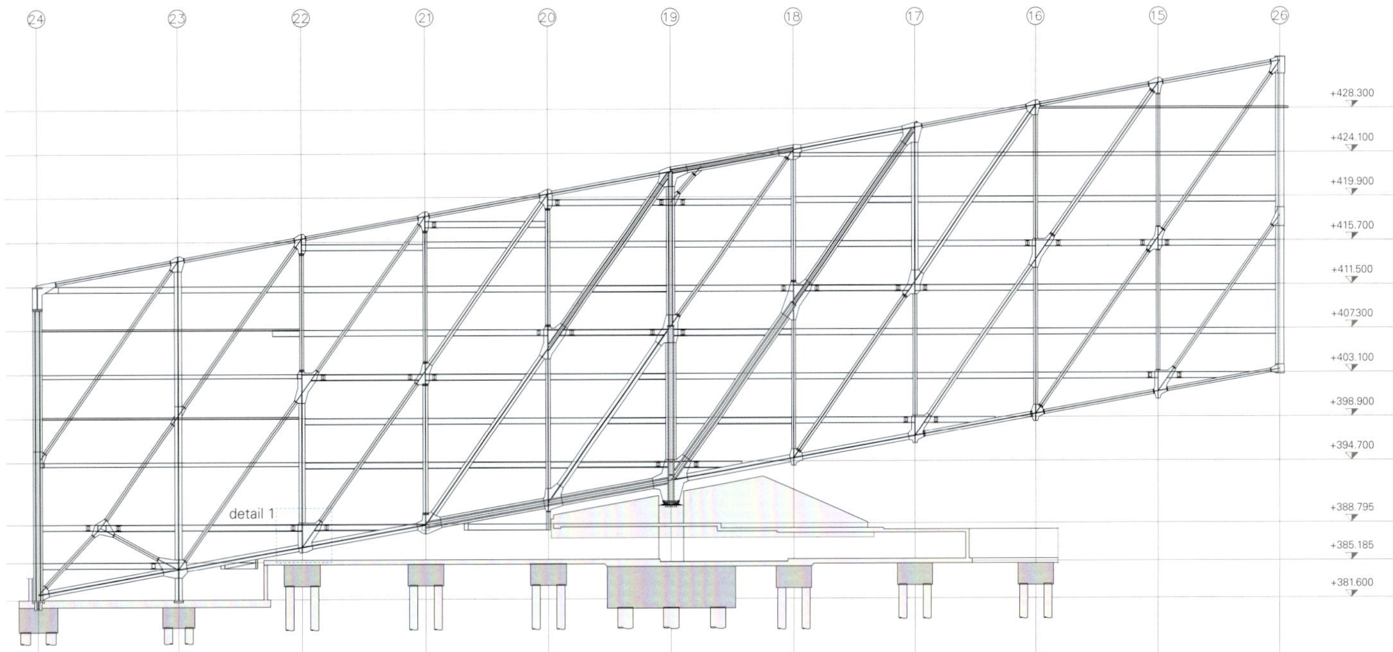

桁架立面 truss elevation

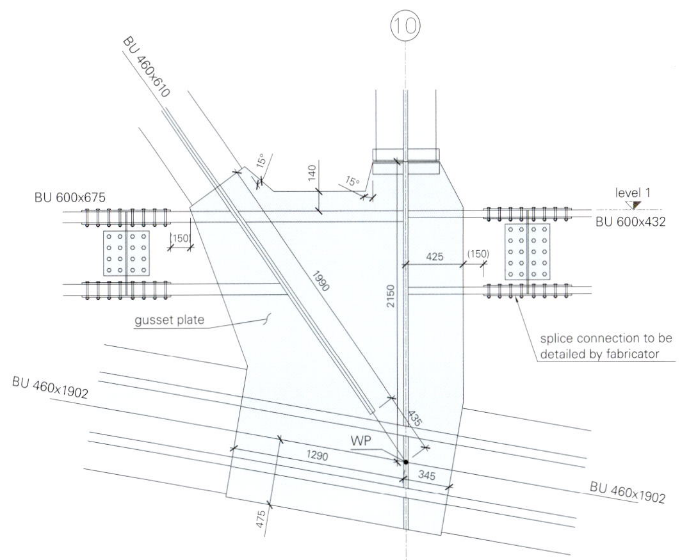

详图1 detail 1

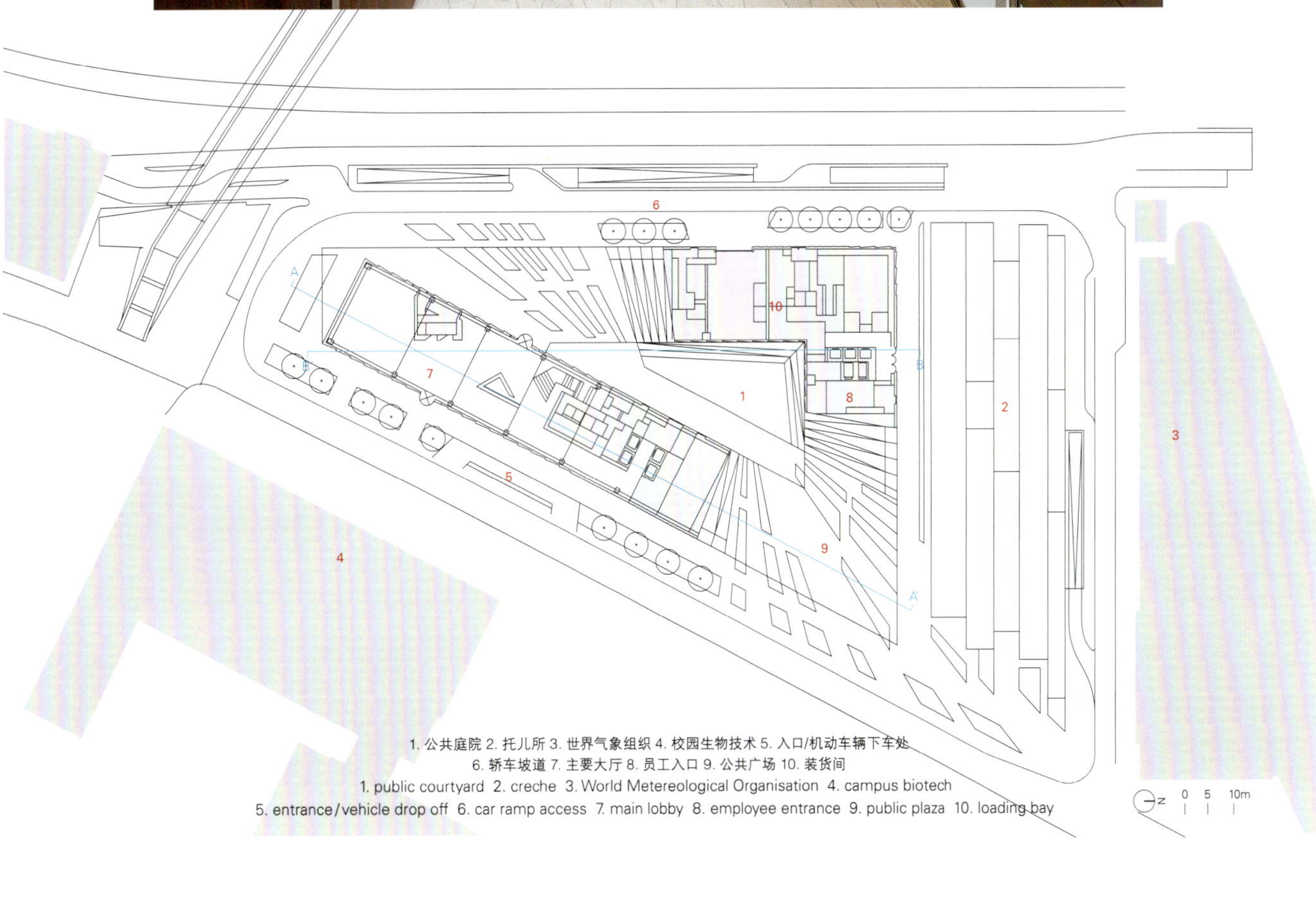

1. 公共庭院 2. 托儿所 3. 世界气象组织 4. 校园生物技术 5. 入口/机动车辆下车处 6. 轿车坡道 7. 主要大厅 8. 员工入口 9. 公共广场 10. 装货间

1. public courtyard 2. creche 3. World Metereological Organisation 4. campus biotech 5. entrance/vehicle drop off 6. car ramp access 7. main lobby 8. employee entrance 9. public plaza 10. loading bay

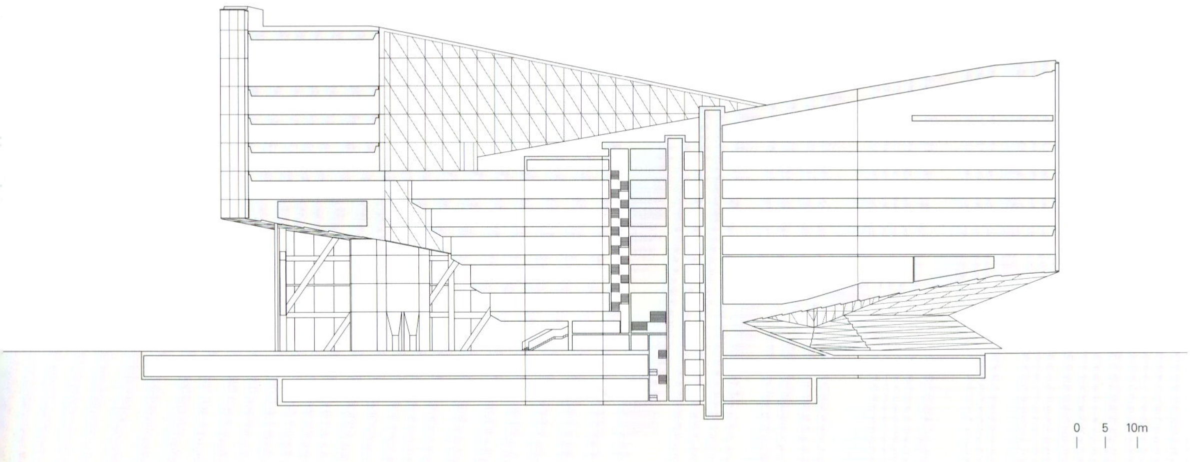

A–A' 剖面图 section A-A'

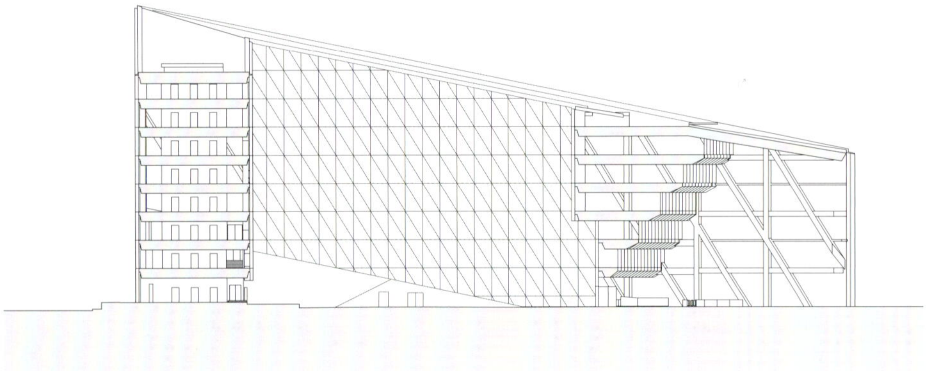

B–B' 剖面图 section B-B'

auditorium, fitness center, 2 roof terraces and a number of social spaces. This pathway is distinguished by the use of a consistent colour and furniture palette employing regionally sourced materials with robust finishes. A series of curated installations by internationally renowned artists Liam Gillick and Sol LeWitt meanwhile serve as social hubs and way-finding tools for connectivity.

The interior design concept maximises both vertical and horizontal interconnectivity, achieving the client's desire to create a social and collaborative community space for its 1,000+ employees. SOM developed tailor-made workplace models for use throughout JTI's headquarter space and the firm was able to break down corporate silos and contribute to the creation of a seamless workplace network, providing long-term flexibility to accommodate JTI's evolving needs.

The open-plan space was achieved using an innovative engineering design solution; a peripheral torsional tube structural steel system allowing for column-free floor plates spanning 18 meters.

The building's innovative Closed Cavity Facade (CCF) was designed in collaboration with Josef Gartner GmbH as a unitised curtain wall system which responds to the demands of seasonally changing external climatic conditions whilst providing exceptional views out and maximising daylight penetration into the workspace. The floor-to-ceiling glazed panels measuring 3m wide x 4.2m high consist of triple glazing on the inner layer and single glazing on the outer, forming a cavity with a fabric roller blind in between. The panels are sealed and equipped with a pressurised supply of filtered and dehumidified air which prevents condensation and heat build-up inside the cavity. The CCF system prioritises occupant comfort and reduces the whole building's energy demand and carbon emissions, helping it to meet the requirements of European Energy Directives and the Swiss Minergie sustainability rating. The system represents one of the best performing all glass facade systems in SOM's history.

The building has been widely recognised for its positive impact on the community, where it is hoped the aspirational design will be a catalyst for future development in the area. The project, which ranks among the most sustainable in Europe, also includes a 1,162 m^2 day-care center for 104 children and 8,162 m^2 of new public space, which was built in collaboration with the City of Geneva.

"Cube 2" 办公大楼

Estudio Carme Pinós

不可否认的是，高楼大厦总是给人一种奇异感，其如雕塑般设计的价值是其最宝贵的东西。

在这个项目中，我们不希望将这种如雕塑般设计的价值与结构表现分离开来；我们认为，所有平衡的设计，都是既经过仔细的测量计算，同时又让人容易理解的；既充满张力，同时又非常和谐。

由于该建筑离两座高度非常相似的大厦非常近，所以建筑师尝试将它向主街倾斜，从而使其从建筑群中脱颖而出。我们的设计想要表达的就是这种平衡的理念。

我们从一个平行六面体着手，两组混凝土隔板形成一个锐角，将建筑从室内柱群中释放出来，被划分成两个三角形空间。

这些隔板是主要的结构元素。为了打造建筑入口，在首层的立面柱体上采用了斜梁。这种梁的几何形态使建筑立面的"百叶窗"变得倾斜，这样的倾斜也使建筑看起来更加细长。就是在这些相互平衡的考量中，我们营造了这样一个向前倾斜、其外立面的线条不与地面平行的大厦形象。

"Cube 2" Office Tower

It can't be denied that a tower has always a will of singularity and that its sculptural value is its most precious asset.

In this case, we don't want to disconnect this sculptural value from its structural expression; we believe that any balancing act which its way of working is easily understood, responding to a carefully measured calculation, will reflect a tension that is, at the same time, harmony.

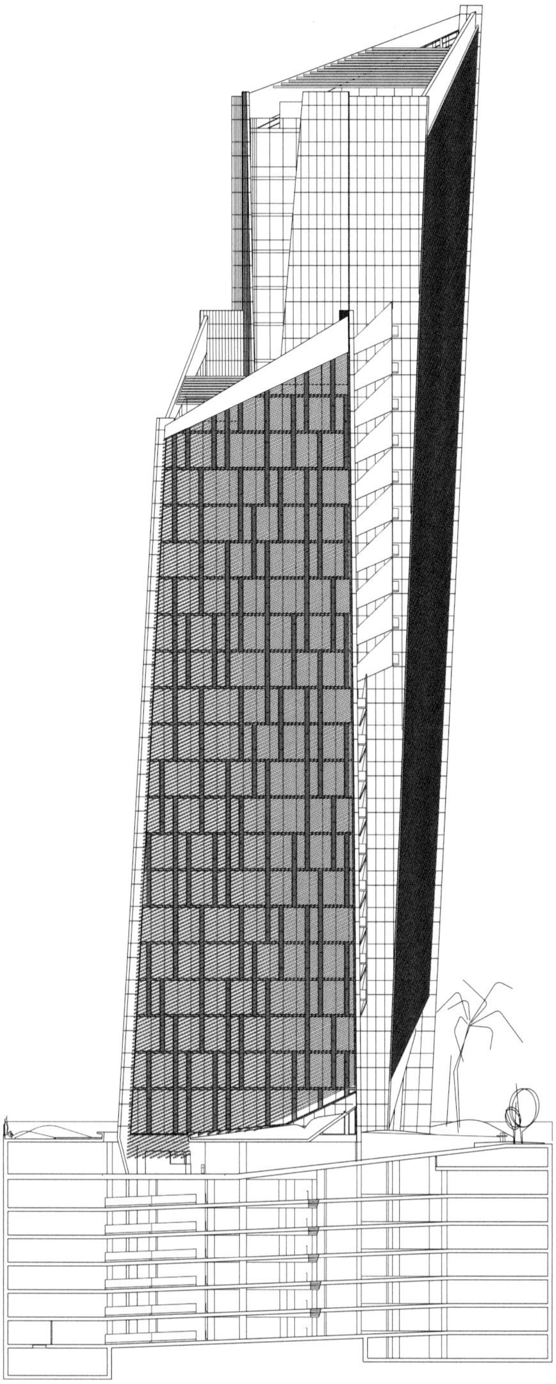

西北立面 north-west elevation

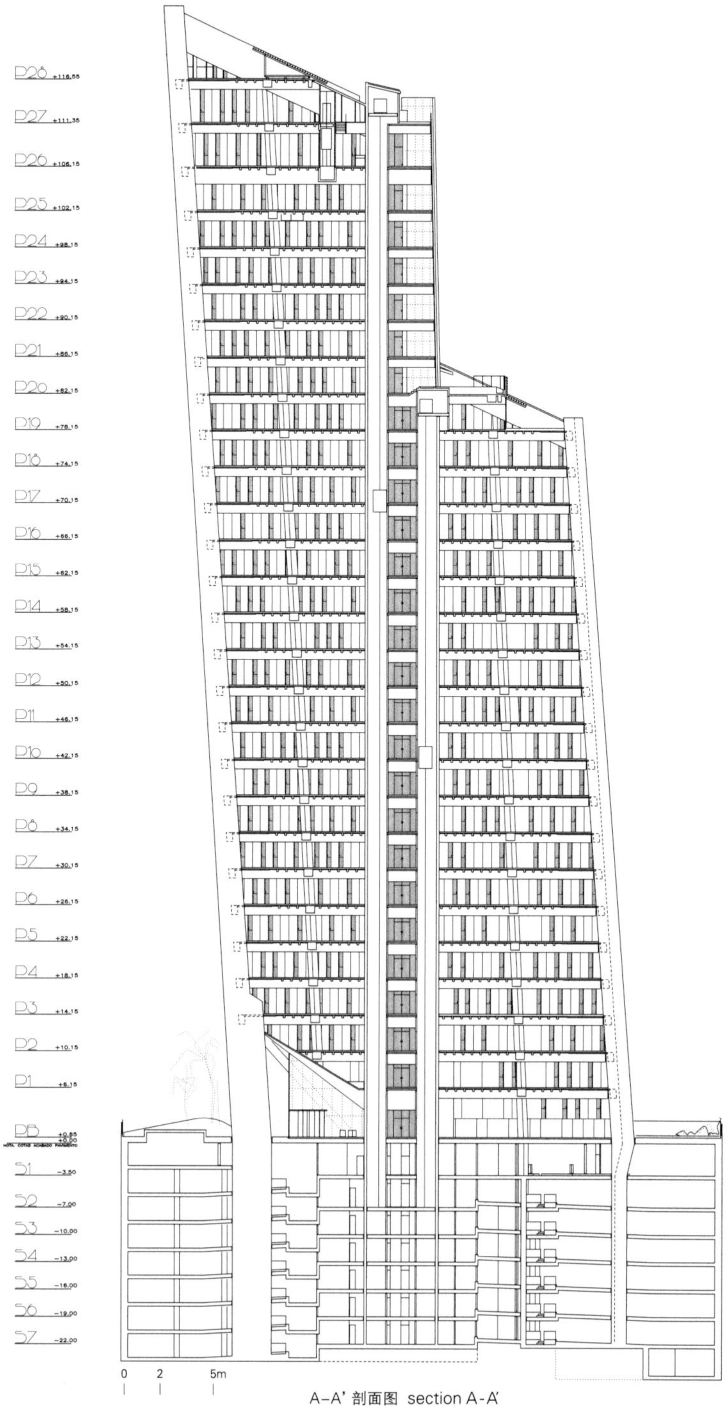

A–A' 剖面图 section A-A'

The building is really close to two towers very similar in height, so we try to make it stand out by inclining it towards the main street, this balancing gesture is the expression of our project.

We start from a parallelepiped shape, two sets of concrete bulkheads form an acute angle and divide the building into 2 triangles freed from interior columns.

These bulkheads are the main structural element. An inclined girder dismounts the facade columns on the ground level in order to make the entrance to the building. This girder's geometry gives the tilting for the facade's "brise-soleil", seeking with this inclination an image of slenderness. Our formal expression is following these balancing games, which are a result of tilting the tower forward, and creating a facade with lines non parallel to the ground. Estudio Carme Pinós

项目名称：Cube 2 Tower / 地点：Paseo royal country, Puerta de Hierro, 45116, Zapopán, Jalisco, México
建筑师：Estudio Carme Pinós / 作者：Carme Pinós
项目团队：Juan Antonio Andreu, Cristobal Martínez, Holger Hennefarth with Daniel Cano, Paula Poveda, Teresa Lluna Andreu, Jose Carlos Castro Barros
模型：MyP Maquetas / 结构：Luis Bozzo / 施工顾问：Salvador Macias, Diego Quirarte
施工团队：Anteus / 客户：Cube Group / 总建筑面积：14,800m^2 / 停车场面积：18,300 m^2
造价：planned budget_USD 19,500,000 / final budget_USD 20,300,000 / 设计时间：2009—2010 / 施工时间：2010—2014
摄影师：©Jordi Bernardó (courtesy of the architect)

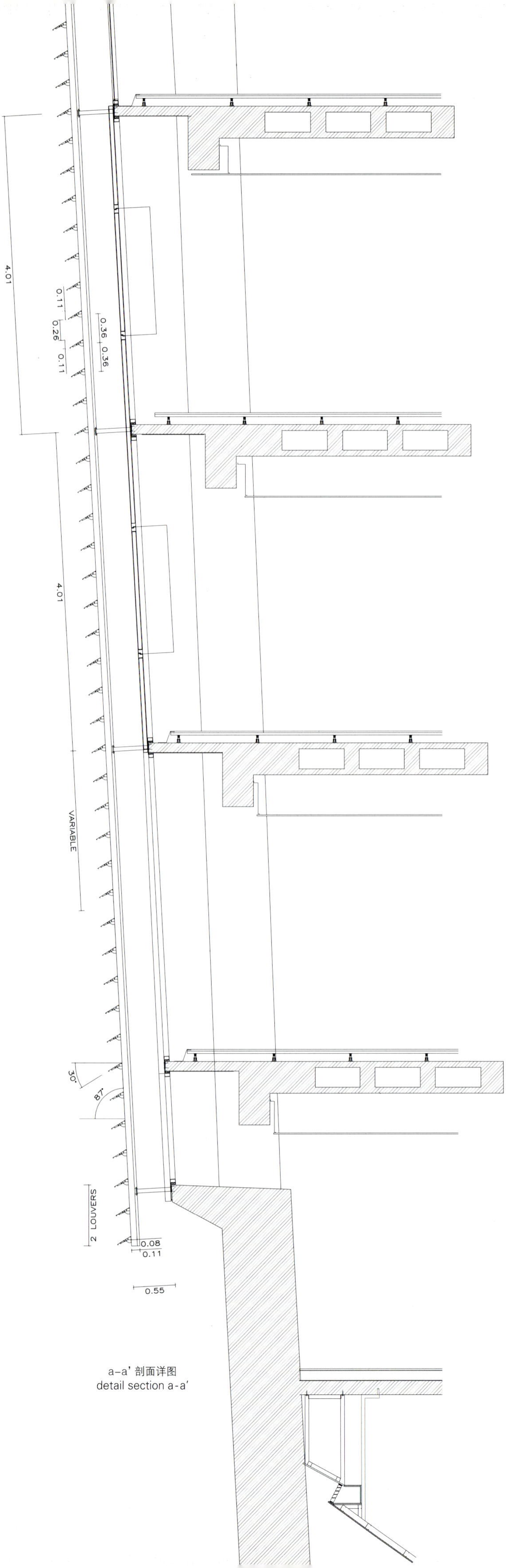

a-a' 剖面详图
detail section a-a'

Sam's
CLUB

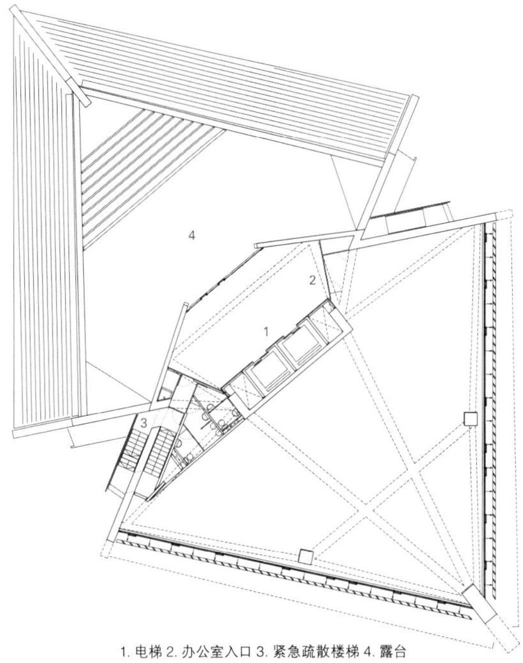

1. 电梯 2. 办公室入口 3. 紧急疏散楼梯 4. 露台
1. lifts 2. entrance to offices 3. emergency staircase 4. terrace
二十二层 twenty first floor

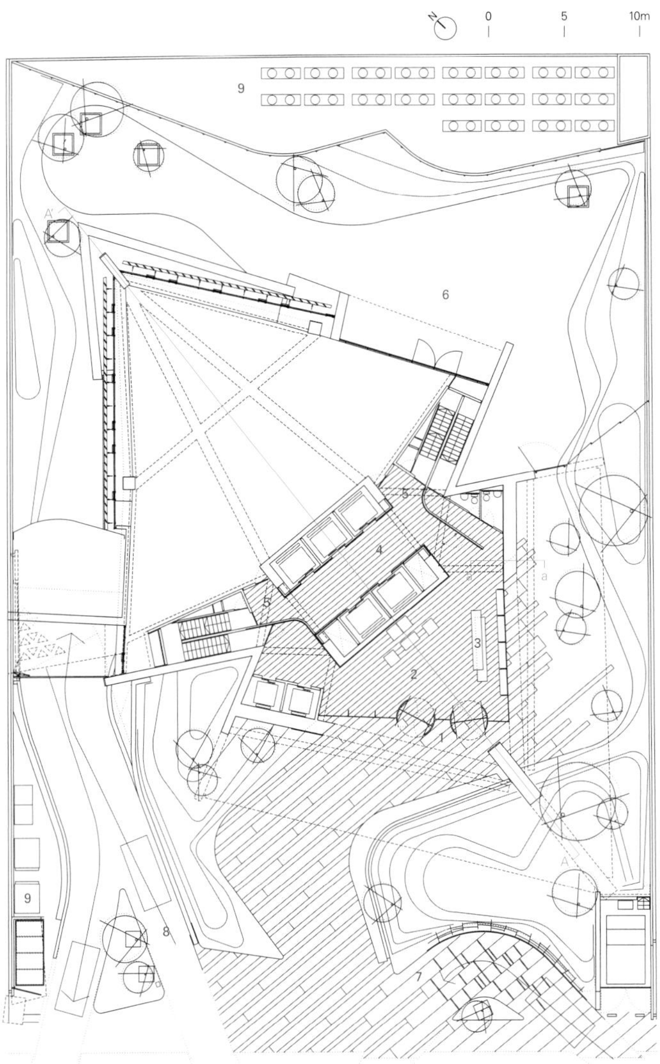

1. 主入口 2. 门厅 3. 接待处 4. 电梯 5. 办公室入口 6. 露台花园 7. 汽车门厅
8. 停车场坡道 9. 设备区
1. main entrance 2. lobby 3. reception 4. lifts 5. entrance to offices
6. terrace garden 7. motor-lobby 8. ramp to parking 9. services
一层 ground floor

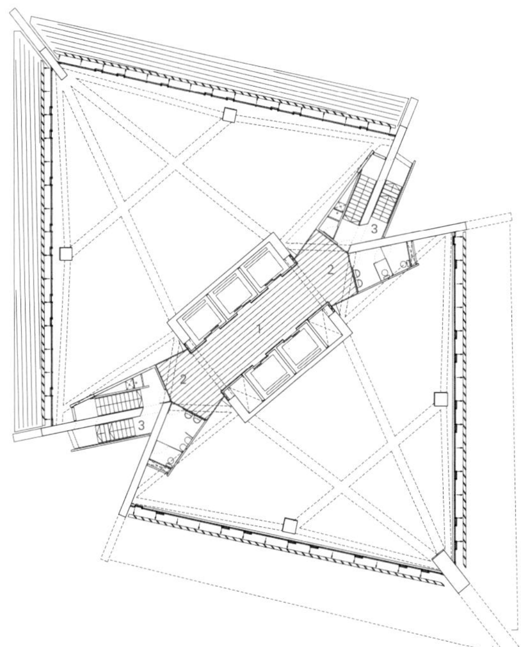

1. 电梯 2. 办公室入口 3. 紧急疏散楼梯
1. lifts 2. entrance to offices 3. emergency staircase
十一层 tenth floor

Bestseller办公综合楼

C.F. Møller Architects

这是一座为800多名员工设计的办公综合楼，它由几栋层高不同的建筑组成，就像一个小型舰队，这些建筑之间通过前庭、院子、露台、屋顶花园等一些室外区域相互连接。

办公楼和室外区域的结合营造出了城中城的感觉。整个办公综合楼被四周的运河和湖泊包围着，形成了进入丹麦阿尔路斯滨水新城区的入口。

几座独立的建筑物层数不同，从一层到十二层不等。综合楼包括办公区、展示厅、摄影与电影工作室、礼堂和宽敞的员工餐厅，再加上可以停放450辆汽车和400辆自行车的三层高停车场，以及低于地面的卸货区。

横向的南北立面有与众不同的凹凸，这是用来自西西里岛的天然石建造的，在立面上围合成了一个个深深的凹槽。这种天然石立面与简单透明的东西两侧的玻璃立面形成了鲜明的对比。室内的楼梯和地板也使用了天然石，甚至包括墙面和家具。它与黑色的橡树和浅色的金属表面相结合，共同营造了色彩和纹理的精致结合。

该办公综合楼使用的是海水冷却系统和太阳能发电，目标是成为2015年度低能耗建筑物之一，即能耗会比建筑规范中的最低要求低50%。

这个综合楼的焦点是有一条内部“街道”以及室内中央广场，位于把建筑物的地面分成了两半的一条通道上方。从广场可以进入公共的礼堂、会议室以及试验的商店区。这个地方可以容纳500到1000人，人们可以聚集到高处、楼梯和阳台上来观看大型时装表演或参加会议。

入口处像是一座桥，从这里可以看到建筑物的地下三层。从内部街道，可以不时地瞥到室外的绿色露台和屋顶花园——这样的绿色室外空间总共有10处。街道直达就餐区，就餐区根据氛围、大小和隐私

程度的不同可以被细分成很多部分。就餐区直通露台，并且穿过原来的2号码头可以到新开发的运河边。

Bestseller Office Complex

The office complex for about 800 employees resembles a varied flotilla of buildings at different levels, which are connected by a series of outdoor spaces like atria, courtyards, terraces and roof gardens.

The mixture of office buildings and outdoor spaces creates the impression that this is a town within a town. The complex, surrounded by canals and lakes on all four sides, forms the entrance to the new, urban district on the waterfront in Aarhus, Denmark.

The individual buildings vary from one to twelve storeys. The complex contains office space, showrooms, photo and film studios, an auditorium and a generous canteen-restaurant area, plus a three-storey car park with space for 450 cars, 400 bicycle racks and goods reception zone below street level.

The latitudinal south and north facades have distinctive reliefs, created in natural stone from Sicily, which frame deep

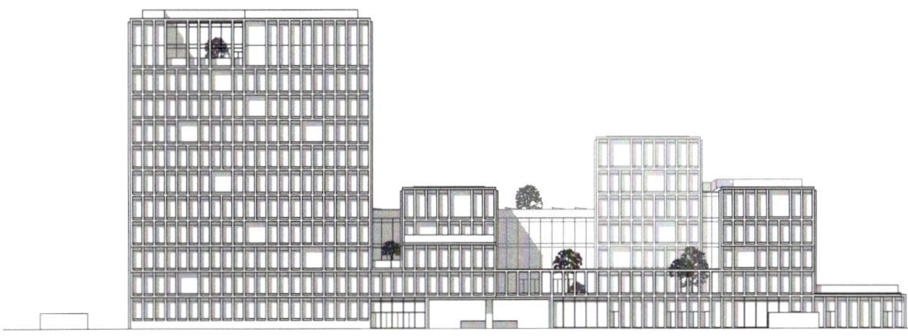

西南立面 south - west elevation

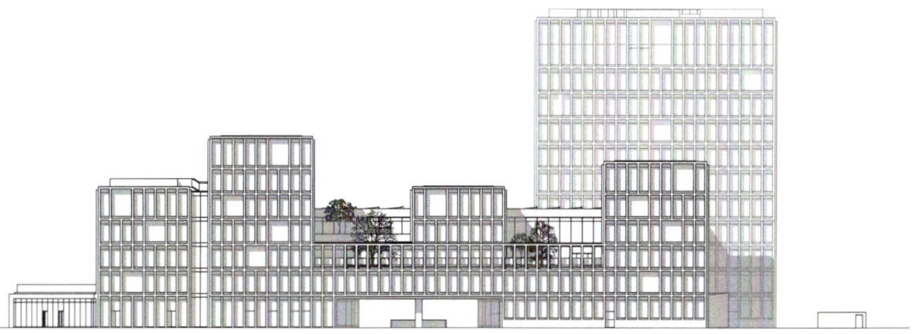

东北立面 north - east elevation

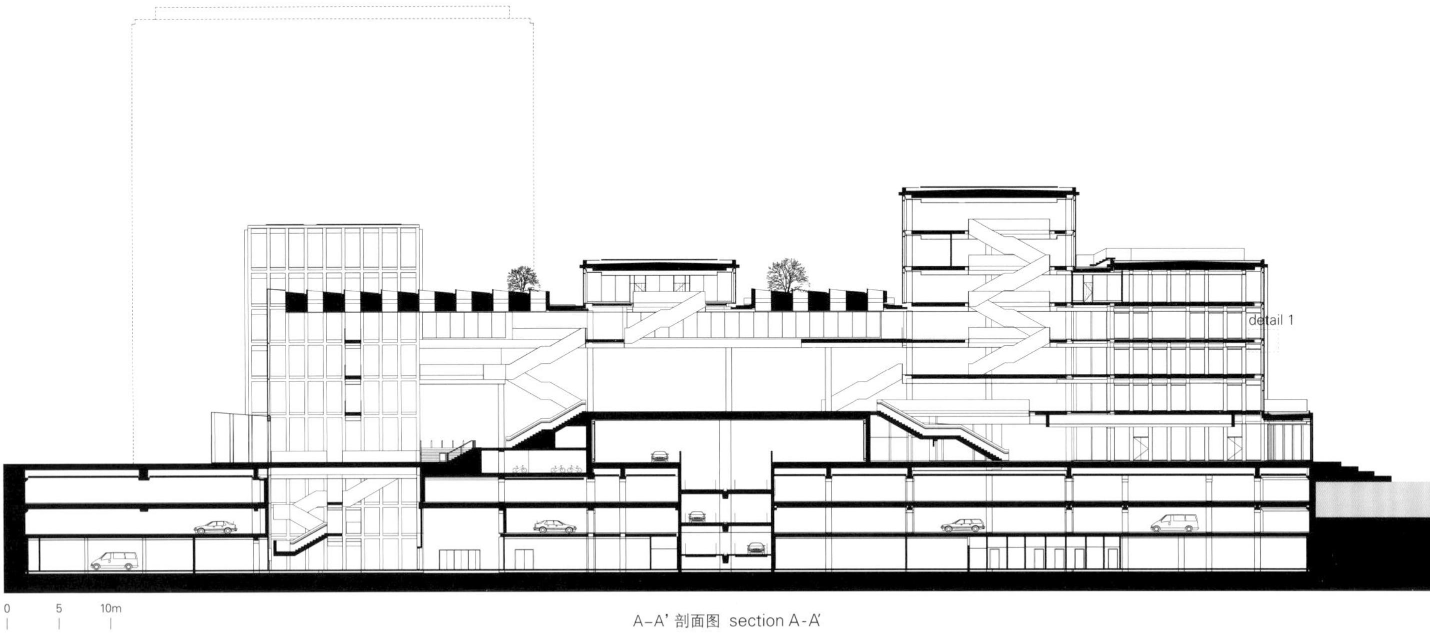

A-A' 剖面图 section A-A'

项目名称：Bestseller office complex
地点：Pier 2, Port of Aarhus, Denmark
建筑师：C.F. Møller Architects
项目团队：C.F. Møller Architects
合作方：C.F. Møller Landscape, COWI (engineering)
景观设计师：C.F. Møller Landscape
客户：Bestseller A/S
用地面积：7,500m²
建筑面积：ca. 5,200m²
总建筑面积：46,000m²
(22,000m² above ground, 24,000m² below ground level)
施工：MT Højgaard, Pihl, Zblin
竣工时间：2015
摄影师：

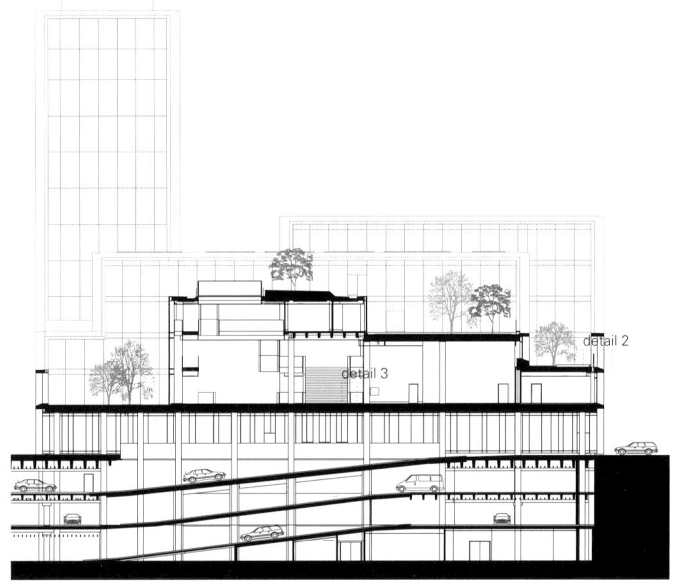

B-B' 剖面图 section B-B'

window niches. These facades contrast sharply with the simple transparent glass facades framed by natural stone to the east and west. Natural stone is also featured indoors – on stairs and floors, and even as wall covering and furniture. It is combined with dark oak and light metal surfaces to create a refined interplay of colours and textures.

The office complex uses sea-water cooling and solar energy and is planned as a low energy class 2015 building, i.e. energy consumption will be 50% lower than the minimum requirements stated in the building regulations.

The focal point of the complex is an internal "street" with a central indoor plaza laid over a road passage which splits the building's ground floor into two halves. From the plaza there is access to a communal auditorium, meeting facilities and experimental store environments. Somewhere between 500 and 1,000 people can gather on the plateau, stairs and balconies to attend large fashion shows and meetings.

The entrance is shaped like a bridge, from which there are views of the building's three underground levels. From the internal street there are varying glimpses into the outer green terraces and rooftop gardens – 10 green outdoor spaces in all. The street leads to the restaurant area, which is sub-divided into many sections with different ambiances, sizes and degrees of privacy, from which there is direct access to terraces and a brand-new canal across the erstwhile Pier 2.

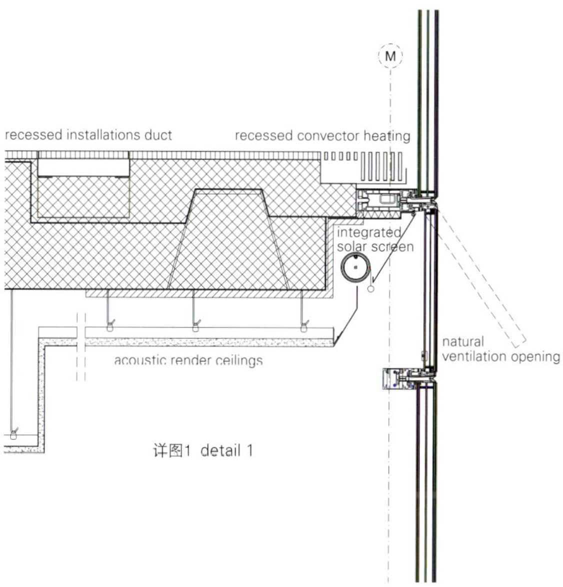

详图1 detail 1

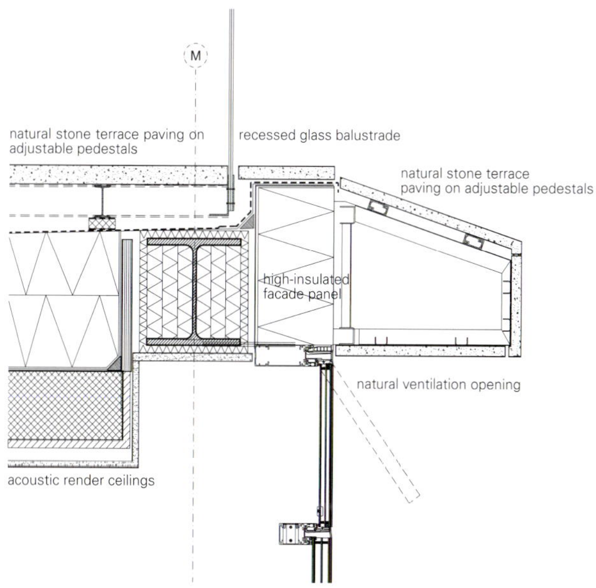

详图2 detail 2

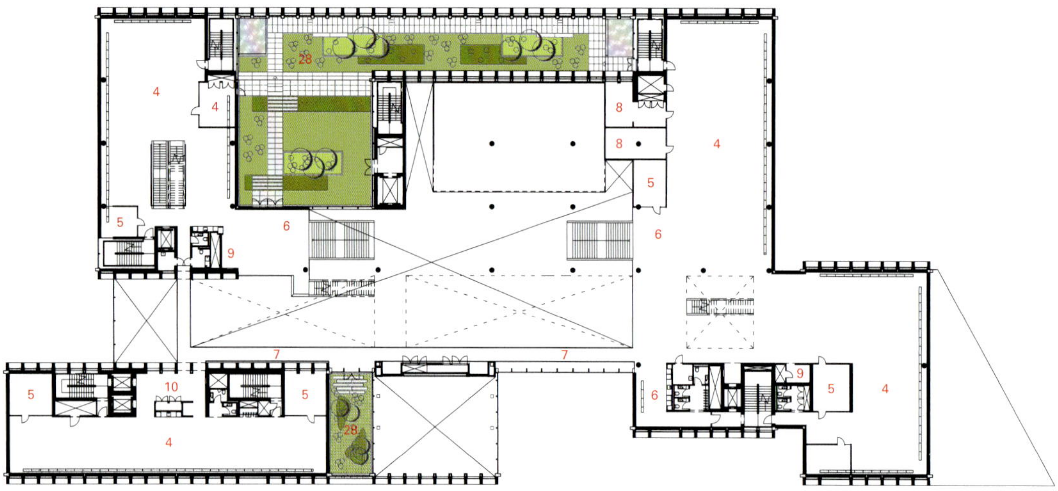

三层 second floor

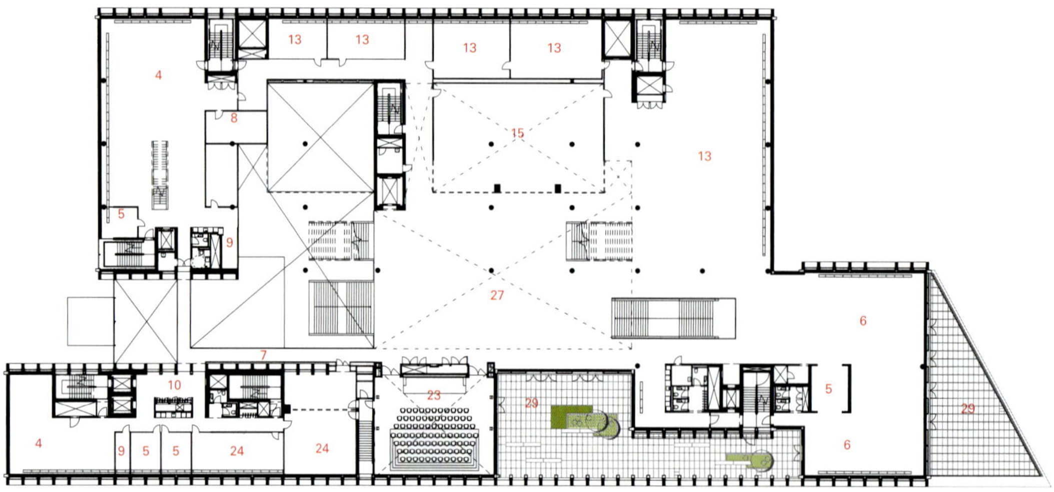

二层 first floor

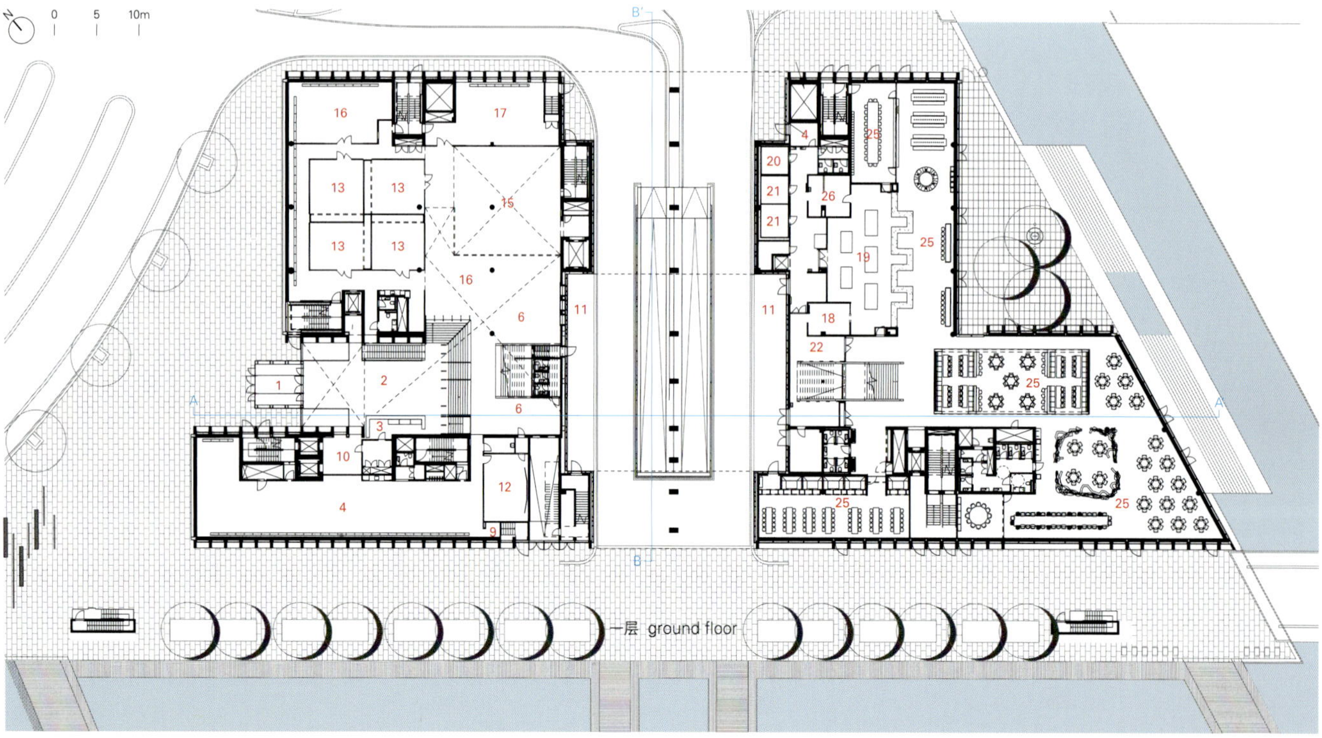

一层 ground floor

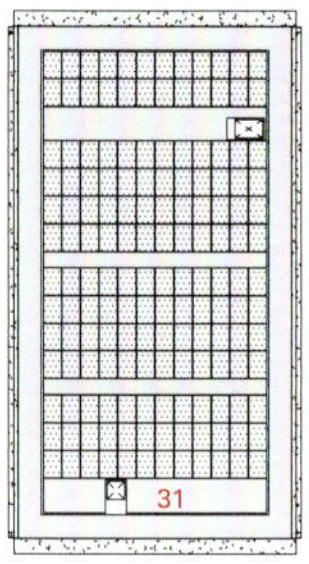

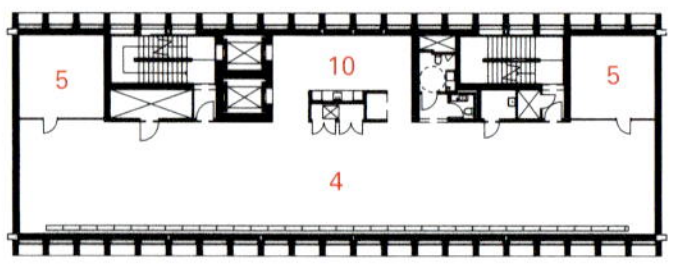

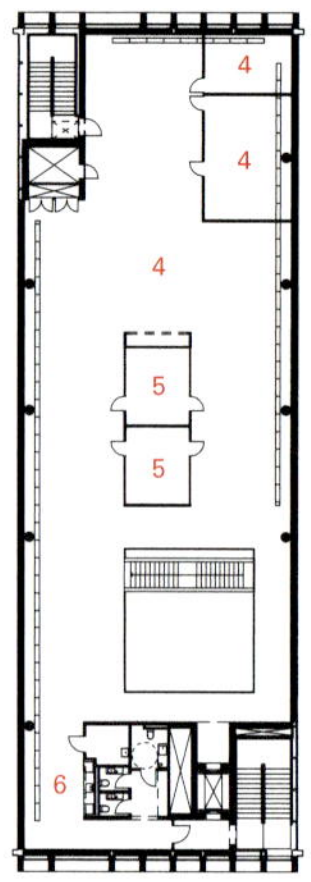

七层 sixth floor

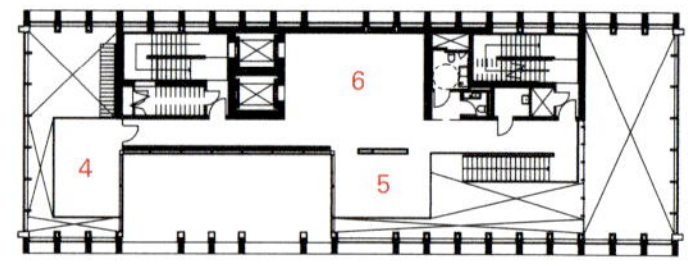

十二层 eleventh floor

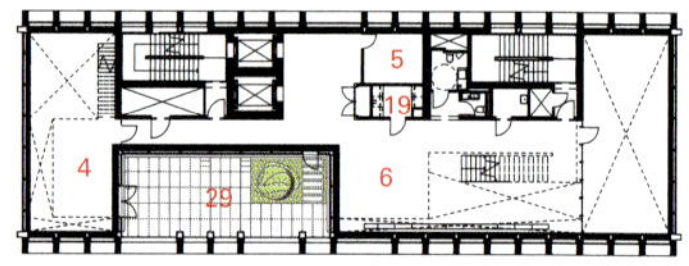

十一层 tenth floor

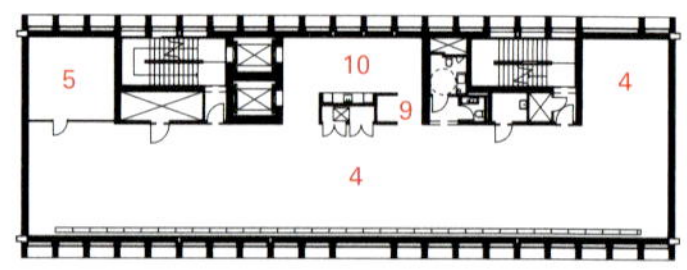

九层 eighth floor

五层 fourth floor

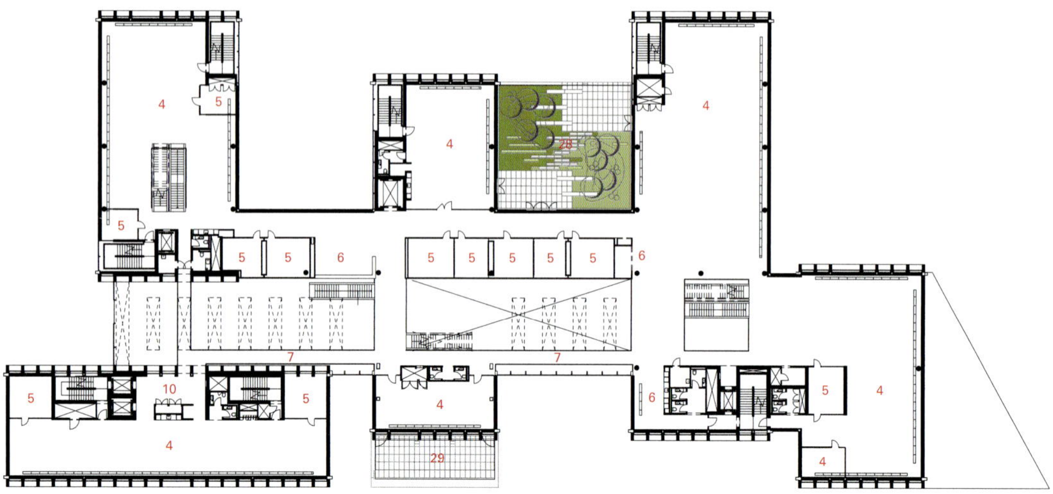

四层 third floor

1. 入口 2. 门厅 3. 接待处 4. 办公室 5. 会议室 6. 休息室 7. 画廊 8. 储藏室 9. 打印室 10. 电梯休息室 11. 卸货区 12. 电视会议室 13. 展示厅 14. 工作坊 15. 店铺展示模型空间 16. 落地工作区 17. 后台 18. 杂货间 19. 厨房 20. 冷冻库 21. 冷却器 22. 洗碗室 23. 礼堂 24. 课堂 25. 餐厅 26. 面包房 27. 论坛 28. 屋顶花园 29. 屋顶露台 30. 摄影工作室 31. 太阳能面板

1. entrance 2. lobby 3. reception 4. office 5. meeting 6. lounge 7. gallery 8. storage 9. print 10. lift-lounge 11. loading bay 12. video conference 13. showroom 14. workshop 15. shop display mock-up space 16. touch down workplaces 17. backstage 18. groceries 19. kitchen 20. freezer 21. cooler 22. dishwashing 23. auditorium 24. tuition 25. restaurant 26. bakery 27. forum 28. roof garden 29. roof terrace 30. photo studio 31. solar panel

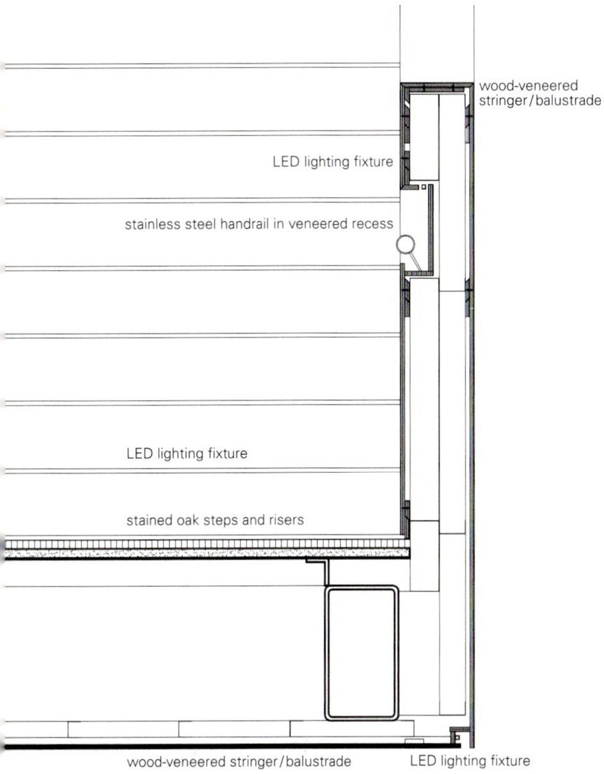

详图2——楼面板
detail 2_floor slabs

©Thomas Moelvig (courtesy of the architect)

Alliander总部

RAU architects

能源电网公司Alliander将其总部改建成了一座不同寻常的可持续建筑和能源产出的综合体，并成为荷兰第一个获得BREEAM-NL杰出可持续认证的改造项目。RAU建筑师事务所负责对原有建筑物进行改造和扩建，以便容纳1550名员工。循环性是本设计不可或缺的组成部分，将建筑改造成材料仓库，材料被暂时存储于其中，而不只是具有有限生命周期的材料的混合体。

现有的综合体由六座不同的建筑物组成，几乎完全保持原样（超过80%的原有表面被保留下来）。从再利用的观点来看，保留原有建筑物的外墙很重要，正因如此，设置了第二层"皮肤"以避免热量损失，降低能源需求。所有外立面原有的窗户都得以保留，而内部的窗户被更换为大面积可打开的窗户，创建出与中庭之间密切的视觉关系，并且形成自然通风。大型中庭上方覆盖着一个标志性的屋顶，在视觉上连接了六个不同的体量，从功能和逻辑上创建了如城市一般的连续空间，促进了员工之间的碰面和交流。屋顶的形状来自于对自然通风的优化研究，因而强化了空气循环。大面积玻璃幕墙和圆形天窗提供了充足的日光，并增强了与景观之间的联系，营造了健康而有启发作用的工作环境。

可持续性被解读为系统的优化：更少的材料、更少的能量以及更少的二氧化碳排放。因此，在设计策略中确立了三个原则：原有材料的保存和再利用、材料使用的最少化，以及材料的利用，以便在以后继续它们的生物或技术生命周期。屋顶的设计是这些策略实现方式的示例。在一家过山车公司的帮助下设计了屋顶的金属结构，实现了更轻盈的结构，减少了原材料的不必要使用，并允许拆卸以供日后再利用。附加在现有体量外侧的新立面使用废弃木柴建造而成，其材料来自于旧电缆盘和当地发现的电线杆。这不仅赋予建筑具有特色和清新的外观，而且也建立了建筑和公司之间的纽带。

该建筑既实用又高效，多出来的能源可重新分配到当地的电网。覆盖在停车场上方的太阳能电池板每年可输电150万千瓦时，并利用地热装置和冷库积累过剩的能量。

由于首先安放了太阳能电池板，这里得以成为荷兰第一个实现正向能源产出的建筑场地。

Alliander总部是荷兰的第一座循环建筑，"材料通行证"是详细列举所有材料种类和数量的文件，无论是在既有建筑还是在新建筑中存在的材料。通过为材料提供身份和充足的信息以防浪费。该文件包含的信息包括处理材料的经手人、临时存放材料的地点以及再利用材料的方式。

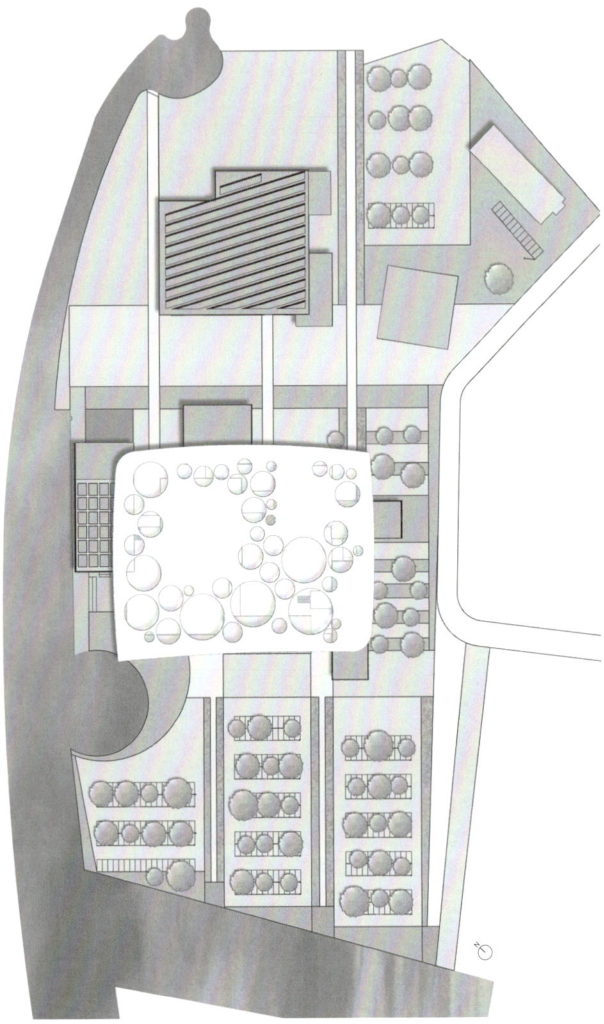

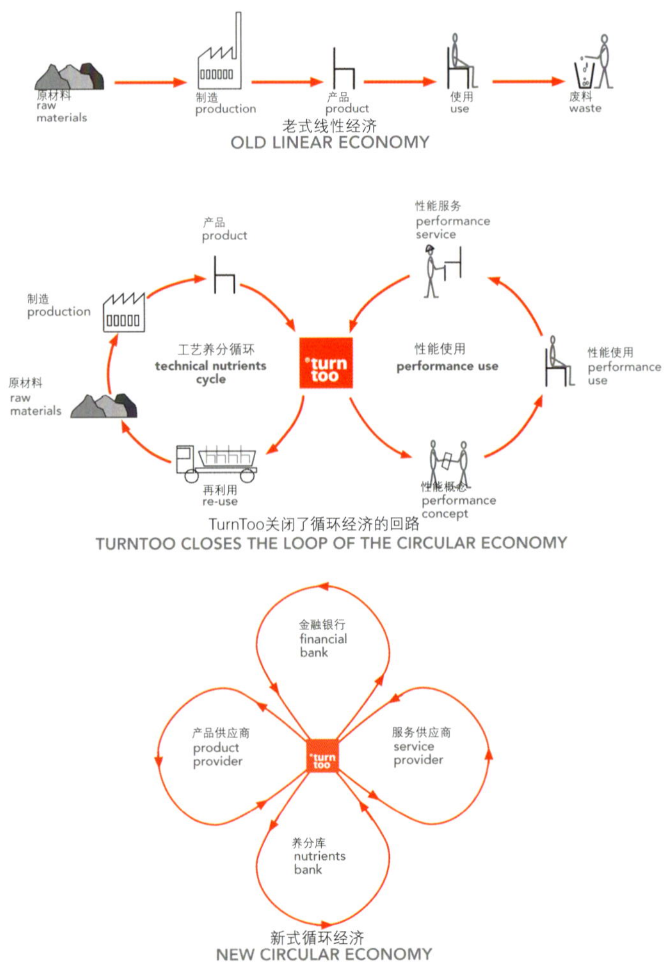

Alliander HQ

The energy grid company Alliander has transformed its headquarters into a remarkably sustainable building and energy positive complex, becoming the first renovation project in the Netherlands to obtain the BREEAM-NL outstanding sustainability certificate. RAU architects have been responsible for the renovation of the existing buildings and the extension, which houses 1550 workers. Circularity has been an integral part of the design, transforming the building into a material depot where materials are temporarily stored rather than just being a conglomeration of materials with a limited life cycle. The existing complex was composed of six different constructions, which are almost entirely maintained (over 80% of the original surfaces remain). Maintaining the facades of the existing buildings was important from a reutilization standpoint, therefore, a second "skin" is placed in order to avoid heat losses, reducing the energy demand. The existing windows are kept in all the facades except the inner ones, where larger operable windows are placed, creating a close visual relation with the atrium and allowing natural ventilation. A large atrium covered by an iconic roof connects the six different volumes visually, programmatically as well as logistically, creating a continuous urban-like space which facilitates encounters and communication among employees.

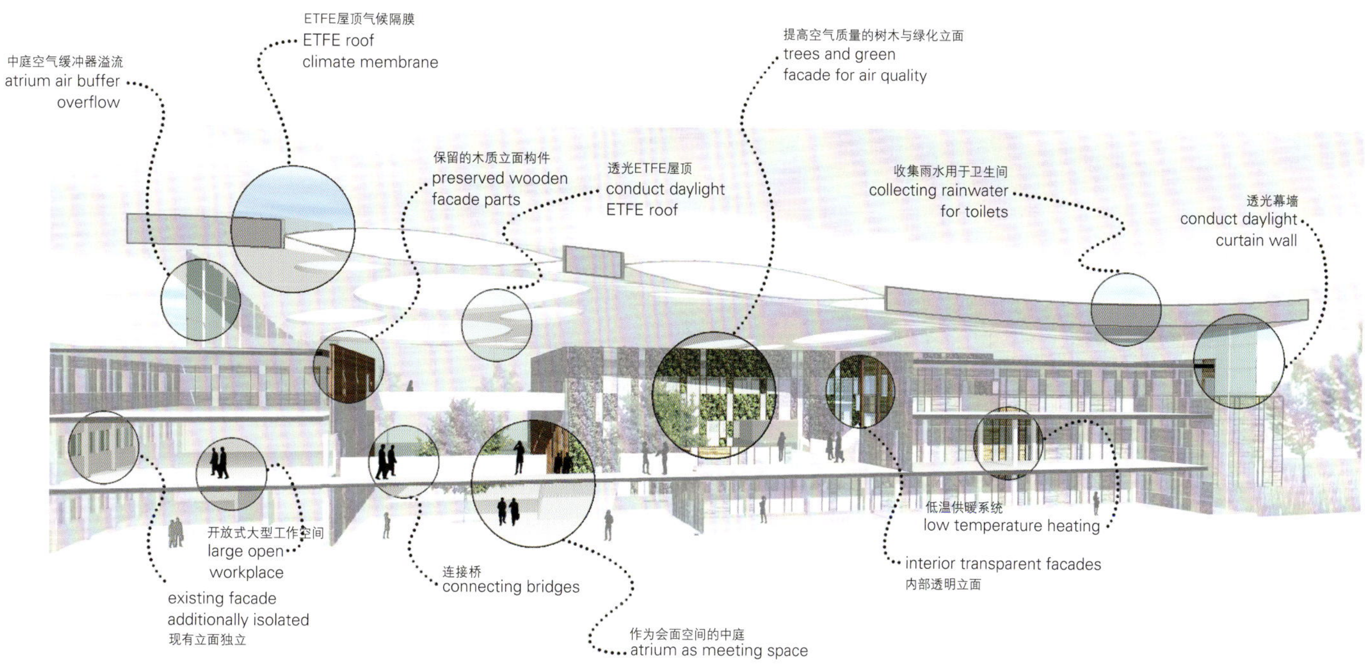

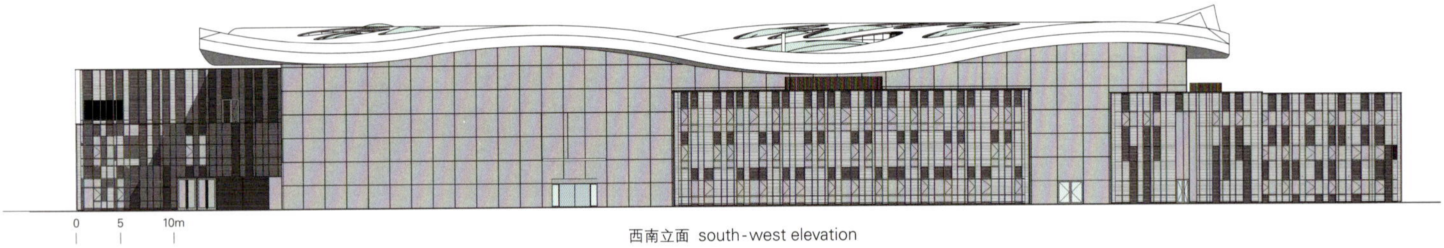

西南立面 south-west elevation

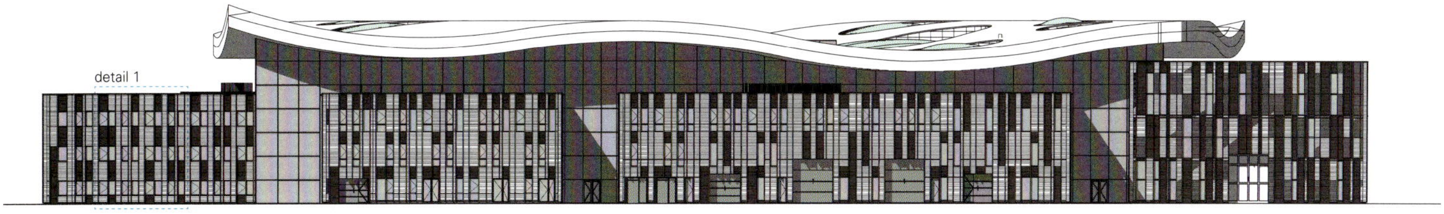

东北立面 north-east elevation

项目名称：Alliander HQ / 地点：Duiven, The Netherlands
建筑师：RAU architects
项目团队：VolkerWessels Vastgoed
合作方/顾问：Fokkema & Partners Architecten _ interior architects, Innax _ installations, Boele & van Eesteren _ contractor, Kuiper Compagnons _ landscape architects, Van Rossum Raadgevende Ingenieurs _ construction, turntoo _ system innovation, Primum _ sustainability advice, EPEA _ C2C advice
用地面积：56,846m² / 建筑面积：14,260m² / 总建筑面积：21,852m²
设计时间：2011—2013 / 施工时间：2013—2015
摄影师：©Marcel van der Burg (courtesy of the architect)

详图1 detail 1

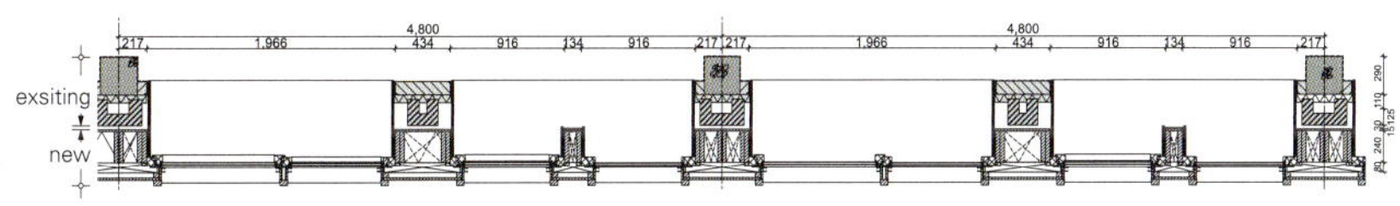

a-a' 剖面图 section a-a'

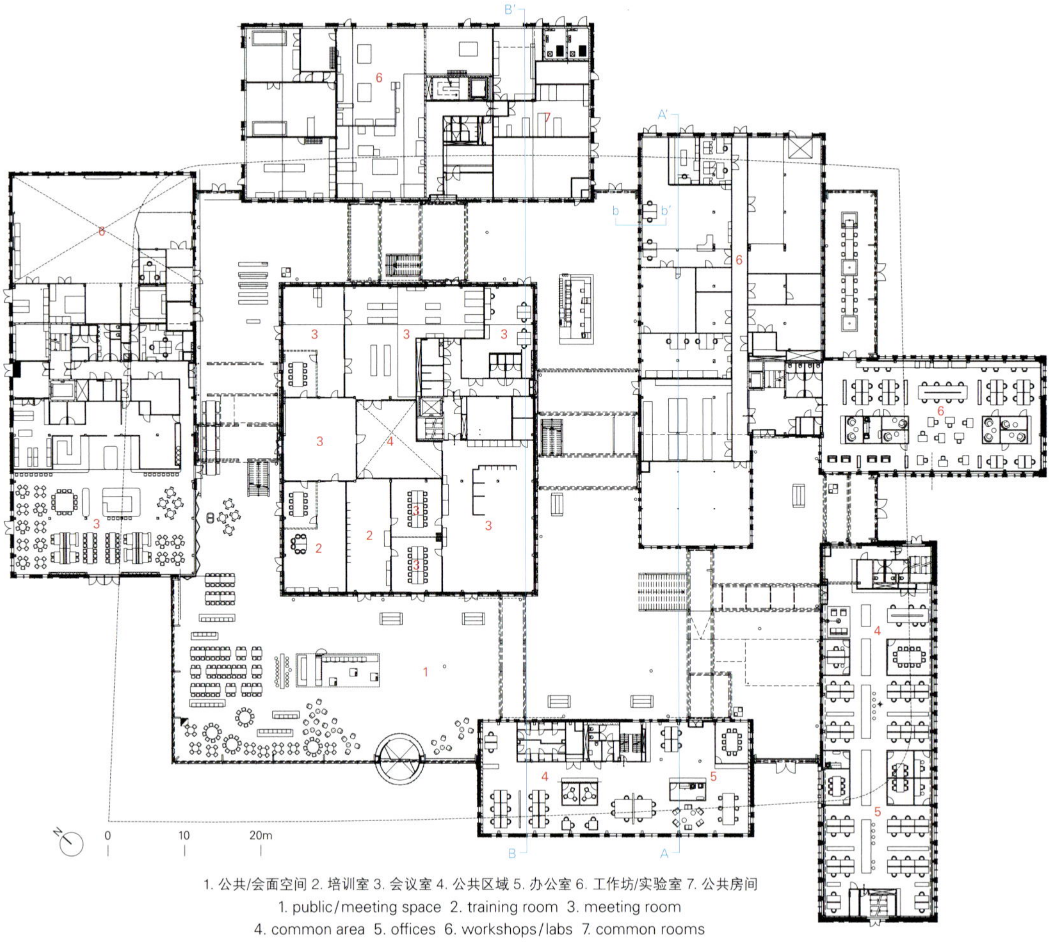

1. 公共/会面空间 2. 培训室 3. 会议室 4. 公共区域 5. 办公室 6. 工作坊/实验室 7. 公共房间
1. public/meeting space 2. training room 3. meeting room
4. common area 5. offices 6. workshops/labs 7. common rooms
一层 ground floor

The shape of the roof is derived from studies for the optimization of natural ventilation, which reinforce the air circulation. The large glass facades and the circular skylights provide the adequate amount of daylight, and strengthen the relation with the landscape, contributing to a healthy and inspiring workspace.

Sustainability is understood as an optimization of the system: fewer materials, less energy, less CO_2. Therefore, three principles were established as design strategies: conservation and reuse of the existing materials, minimization of material use, and employment of materials that can later on continue their biological or technical life cycle. The design of the roof was an example of how these strategies were implemented. A roller coaster company helped with the design of the metal structure, achieving a lighter construction, reducing the unnecessary use of raw materials and allowing disassembly for later reuse. The new facades attached to the existing volumes are built using scrap wood that comes from the old cable coils and utility poles found in the terrain. This not only gives a characteristic and fresh appearance to the building but also establishes a bond between the building and the company. The building is both effective and efficient, providing an energy surplus which can be redistributed into the local grid. The solar panels, covering the parking spaces deliver 1.5 million kWh annually, and the ground heat/cold storage accumulates the excess.

Since the solar panels were placed first, it was possible to achieve an energy positive building site for the first time in the Netherlands.

Alliander is the first circular building in the Netherlands, and the "material passport" is the document that specifies the types and amounts of all the materials present in the building, both preexisting and new. By providing materials with an identity and adequate information, waste is prevented. The document includes information about who has handled the materials, where they are temporarily stored and ways in which they can be reused.

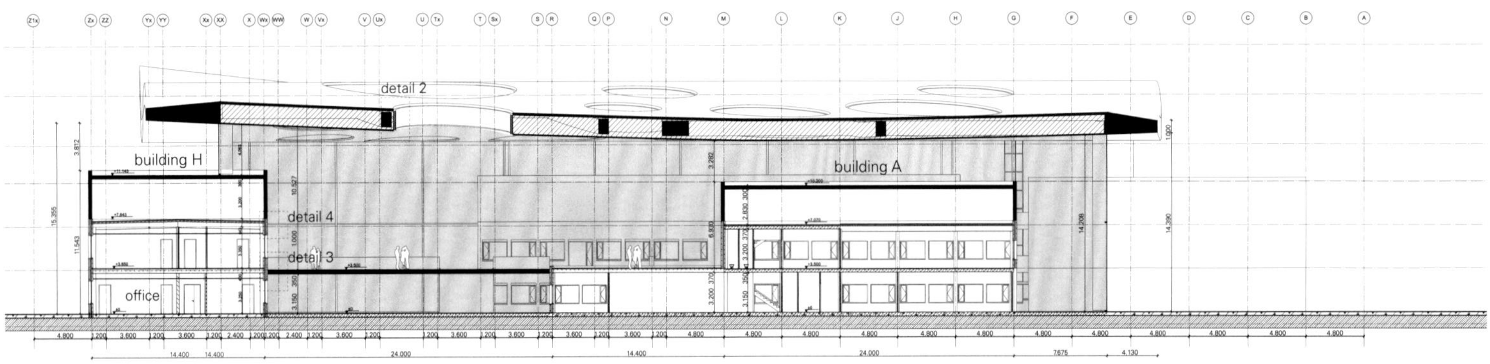

A-A' 剖面图 section A-A'

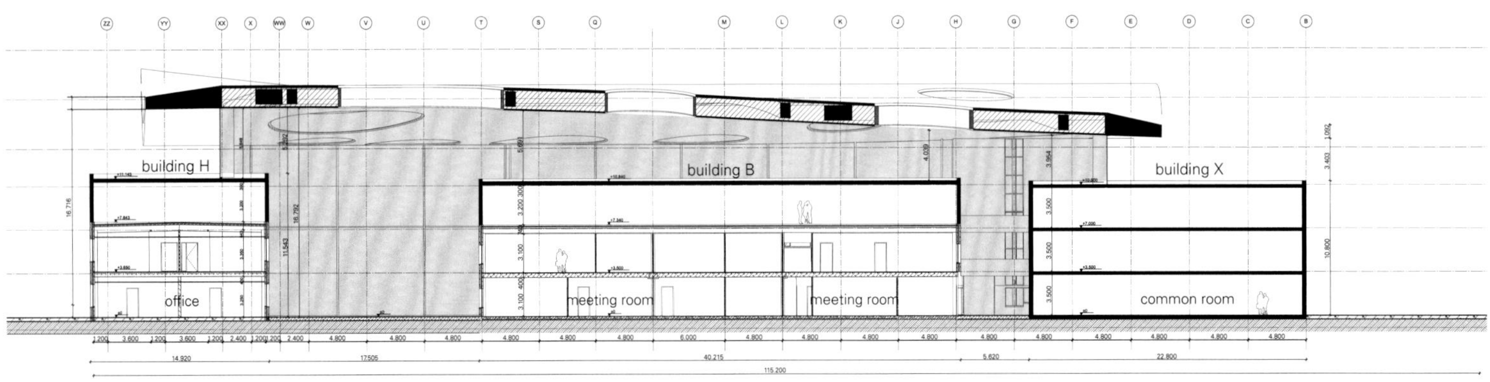

B-B' 剖面图 section B-B'

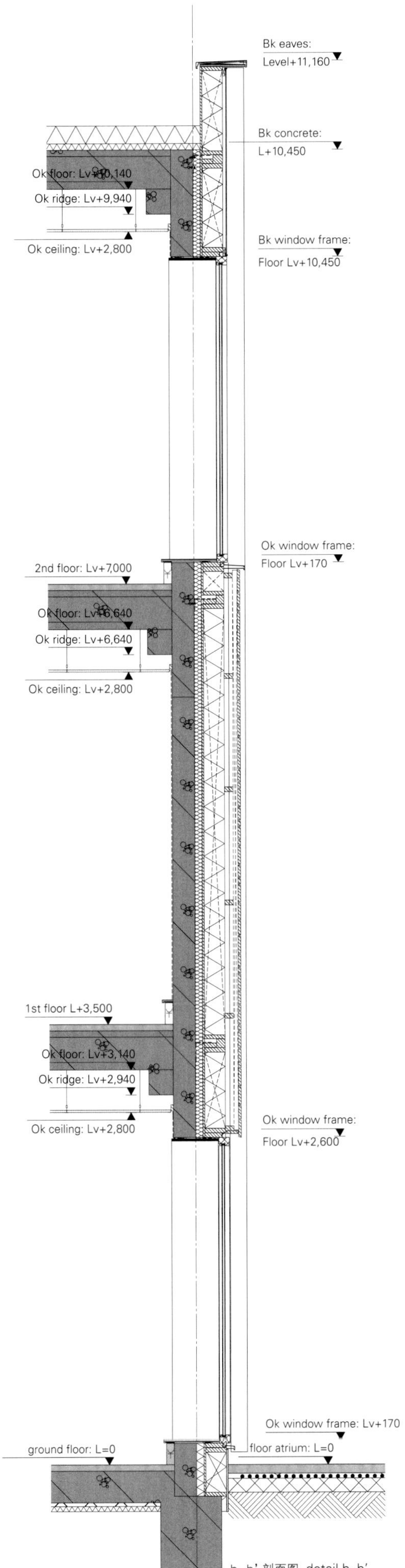

b–b' 剖面图 detail b-b'

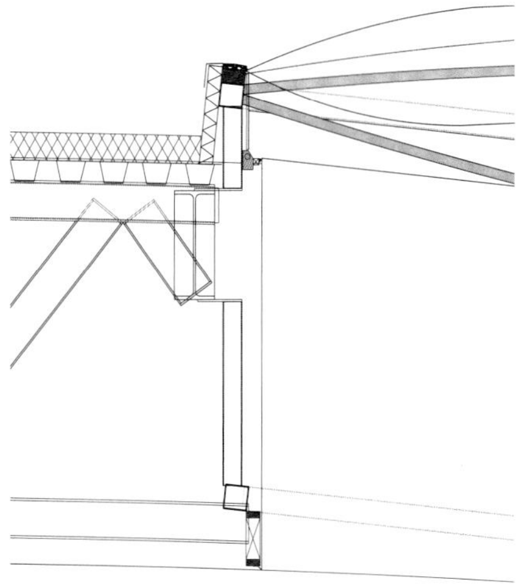

详图2 detail 2

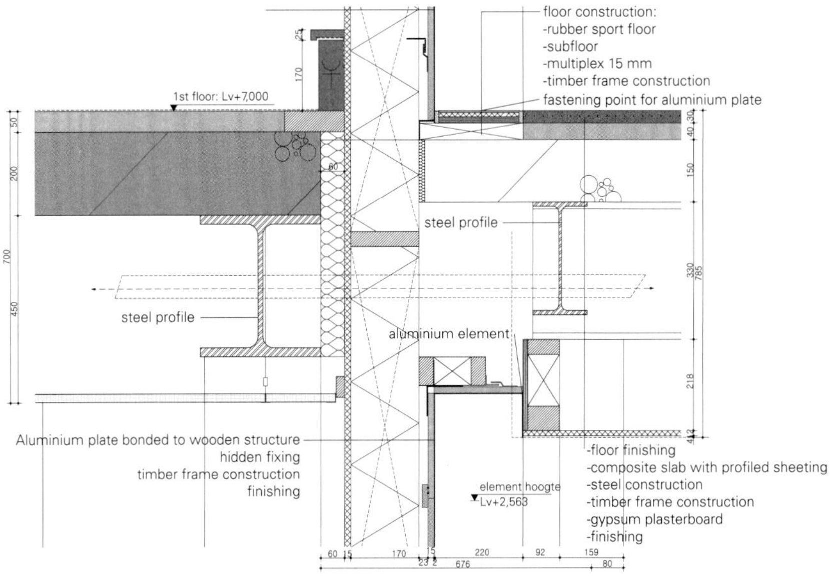

详图3 detail 3

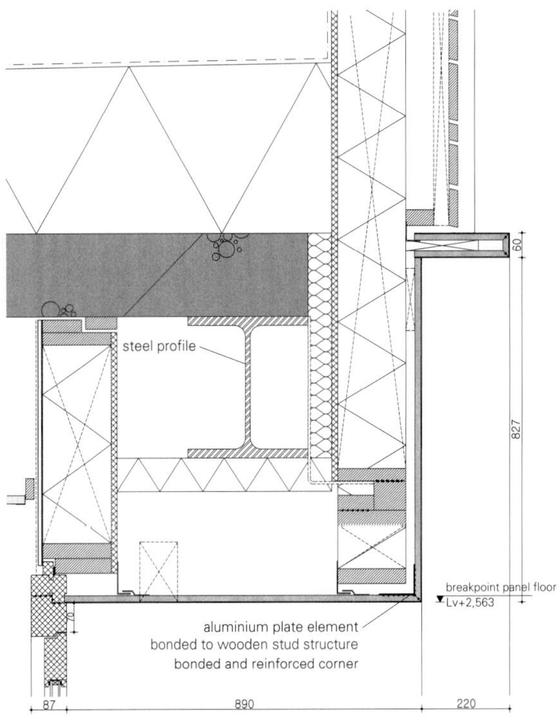

详图4 detail 4

上海仪电集团办公综合楼

Jacques Ferrier Architecture

N
0
10
30m

在2011年2月，雅克·费尔叶建筑设计公司赢得了由上市房地产开发商、上海广电房地产公司举办的办公综合楼和商场的设计大赛，该项目位于上海西南的徐汇区，沿着Shang Ao Dang运河而建。周围社区因为在建地铁，所以一直在扩建。该项目已于2015年夏季竣工。

这个项目由几栋建筑构成。该建筑群的主要特征体现在折叠和切割的相互作用上，这样的设计产生了各种各样的观景视角、视线、光线和不同的建筑高度。这种多样性在提供了灵活的功能性的同时也创造了强烈的画面感。该建筑群主要是南北朝向，这样增强了光线效果，也意味着从大多数办公室都能欣赏到运河的风景。许多区域还设有凉廊，可以看到运河或庭院。

我们提出了"城市地形"的概念，从而打造出"人工景观"的效果，以此来实现建筑与大自然的交融。

该场地位于运河旁，之前隐藏于人们的视野之前。我们的设计提议中最关键的一条是用种植路堤来代替运河沿岸的灰墙。绿树成荫的堤岸现在是运河畔的一条林荫人行道。在这里，大自然重新回归到城市核心的地位上。

办公楼的基础设施也可以被看作景观，绿荫下的凹凸部分朝向地面，就像在邀请人们到来一样。当地居民甚至可以沿着小路走到一楼的屋顶。基础设施对空间进行了非常清晰的划分，赋予这个办公楼项目以个性。运河旁这个景观式基础设施的立面高度各不相同，可以在其中举办各式各样的活动，这里与办公区或独立商铺相连，打造了生机勃勃的公共空间。

景观式的设计大力倡导渗透性，以解决环境问题。该场地超过三分之一的表面都是绿色设计，这并不包括办公楼的基础设施，基础部分的景观设计也可以被称为花园。

该建筑的整体立面采用垂直叶片覆盖的模式，这能保护其不受到上海强烈的太阳光线的破坏。建筑师在这里使用了两种不同的建筑材料：一面被浅金色的铝板所覆盖，另外一面是绿色和蓝色的瓷砖，这是使用古老的技术生产而成的中国传统材料。色彩和材料的相互融合创造出了一种动感，表达了城市建筑不断更新的愿景。根据他们所在位置，人们会看到办公综合楼依次呈现出绿色、蓝色和金色。

Yidian Office Complex

In February 2011, Jacques Ferrier Architecture's firm won a competition held by public real estate developer Shanghai Guang Dian Real Estate for the creation of an office complex and stores along the Shang Ao Dang canal, located in the Xuhui district of south western Shanghai. This neighbourhood has been expanding since the metro line was built. The project was delivered in the summer of 2015.

The project comprises a collection of several buildings. The buildings' identity is defined by an interplay of folds and cuts, which give a variety of perspectives, views, light and heights. This variety provides functional flexibility while creating a strong image. The buildings are mainly north / south facing, which enhances the light and means that most offices offer a view over the canal. Many also feature loggias with views of the canal or the court.

We have developed an urban topography concept to create a "built landscape" effect, where architecture and nature are fused and blended.

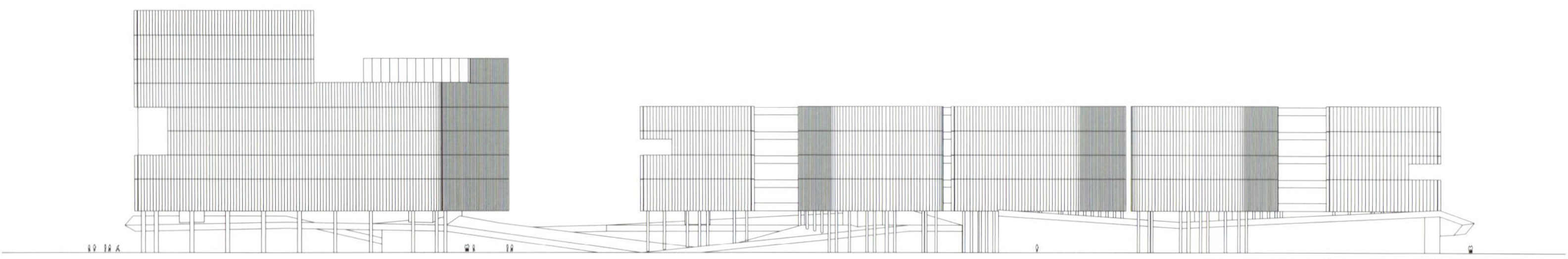

东立面 east elevation

0 5 10m

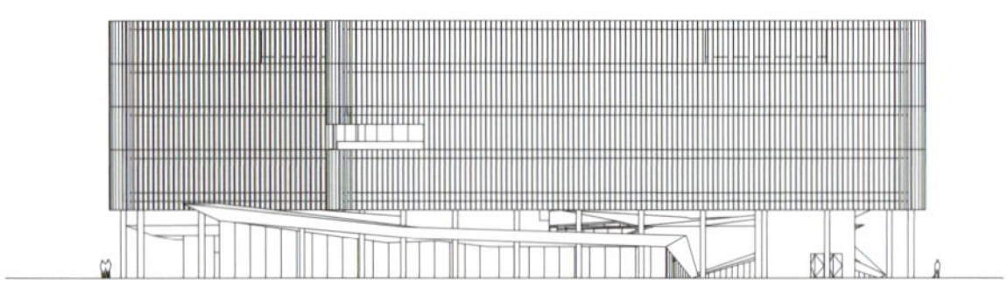

北立面 north elevation

项目名称：Yidian Office Complex / 地点：200 Tianlin Road, Xuhui District, Shanghai, China / 建筑师：Jacques Ferrier Architecture / 合作方：Architectural Design & Research Institute of Tongji University (Group) Co.Ltd. / 结构技术、流体技术设计与经济顾问：idem / 立面：VSA / 景观设计师：DLC / 客户：Shanghai Guang Dian Real Estate / 用途：offices, stores / 建筑面积：60,000m^2 / 造价：EUR 50 M incl. VAT (2011 value) / 设计时间：2011 (competition winning) / 竣工时间：2015.6

摄影师：©Luc Boegly (courtesy of the architect) - p.132~133, p.135, p.136, p.137, p.138~139, p.140~141 / ©Jonathan Leijonhufvud (courtesy of the architect) - p.130~131

一层 ground floor

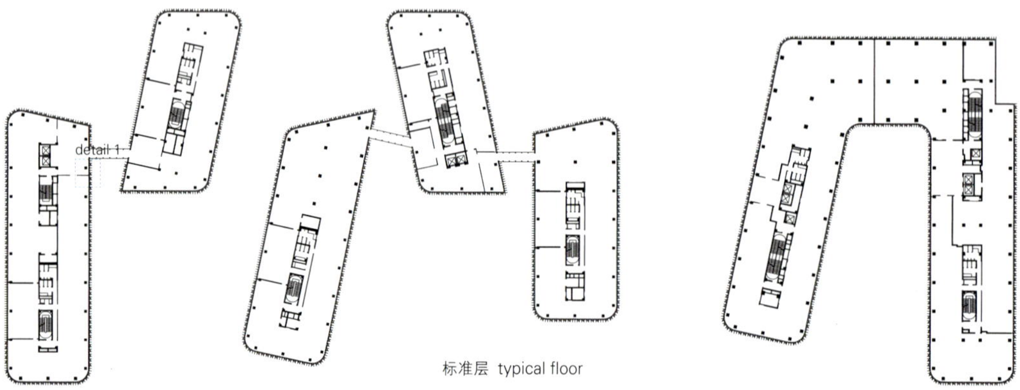

标准层 typical floor

1. 服务设备区入口 2. 服务设备区 3. 研究实验室入口 4. 大厅 5. 火警与保安设备 6. 技术室 7. 停车场出入口
8. 餐厅入口 9. 餐厅 10. 垃圾房设备区 11. 垃圾房办公室 12. 厨房 13. 轿车与自行车停车场 14. 屋顶花园入口
1. entrance to services 2.services 3. entrance to research laboratory 4. hall 5. fire alarm and security
6. technical room 7. entrance-exit to parkings 8. entrance to restaurant 9. restaurant 10. trash room services
11. trash room offices 12. kitchen 13. cars & bicycles 14. entrance to roof garden

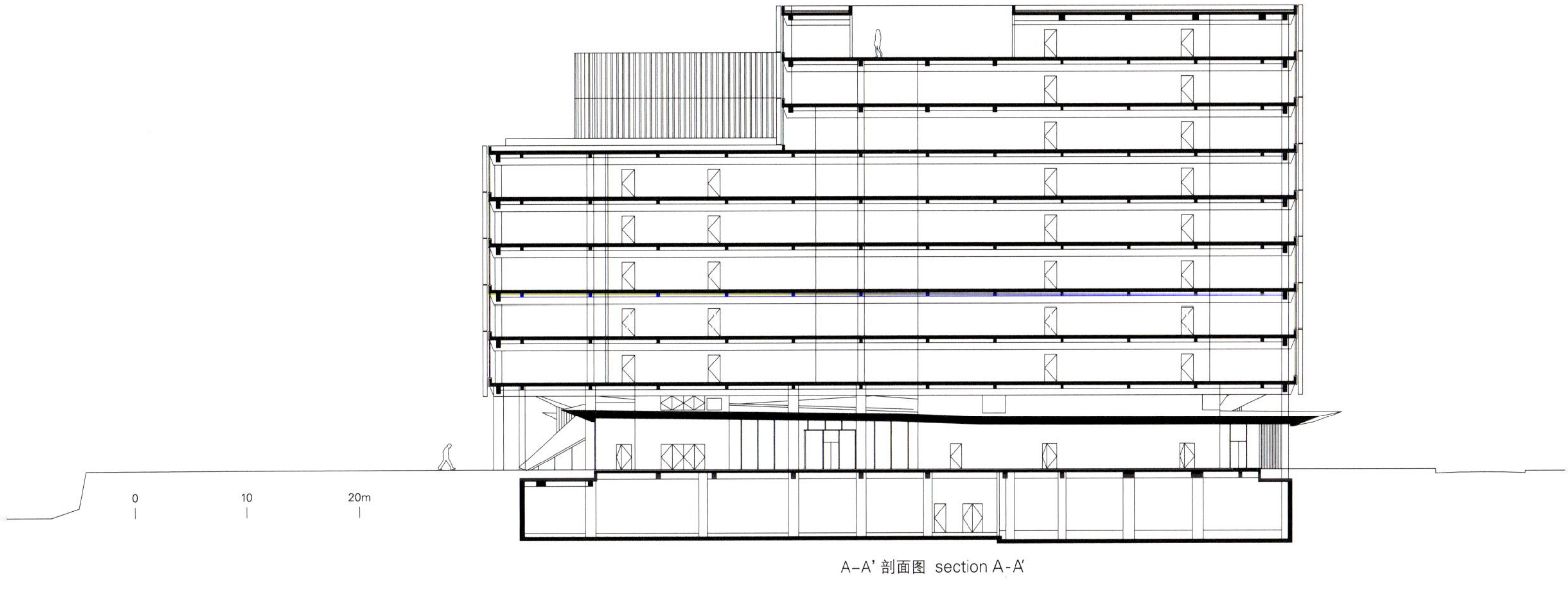

A-A' 剖面图 section A-A'

The site is located by the edge of a canal, which was previously hidden away from the public eye. One of the key points of our proposal is replacing the grey walls along the canal with a planted embankment. The tree-lined embankment is now a public walking path along the canal. Here, nature reclaims its rightful place in the heart of the city.

The office foundation is also treated as a landscape, in which a tree-covered relief with some parts turns towards the ground like an invitation to run across it. Local residents can even walk along a path on the roof of the ground floor. This foundation creates very clear divisions and gives an individual identity to the office space project. Next to the canal, the facade of this landscape-esque foundation of varying heights houses a number of activities, which can be linked to the offices or independent shops, creating lively public spaces.

This landscape-type design encourages permeability, in

response to environmental concerns. More than a third of the site's surface area is dedicated to green spaces, not including the foundation of the office structure, which is also treated as a garden.

The ensemble's facades are covered in vertical blades which protect the structures from the powerful Shanghai sun. They combine two different materials: one side is covered with pale gold aluminium, and the other with ceramic in shades of green and blue, depending on the facades, a traditional Chinese material created using ancient techniques. This dual interplay of colours and materials creates a sense of movement, a constantly renewed vision of the buildings within the city. Depending on their location, urbanites will see the complex as successively green, blue and golden.

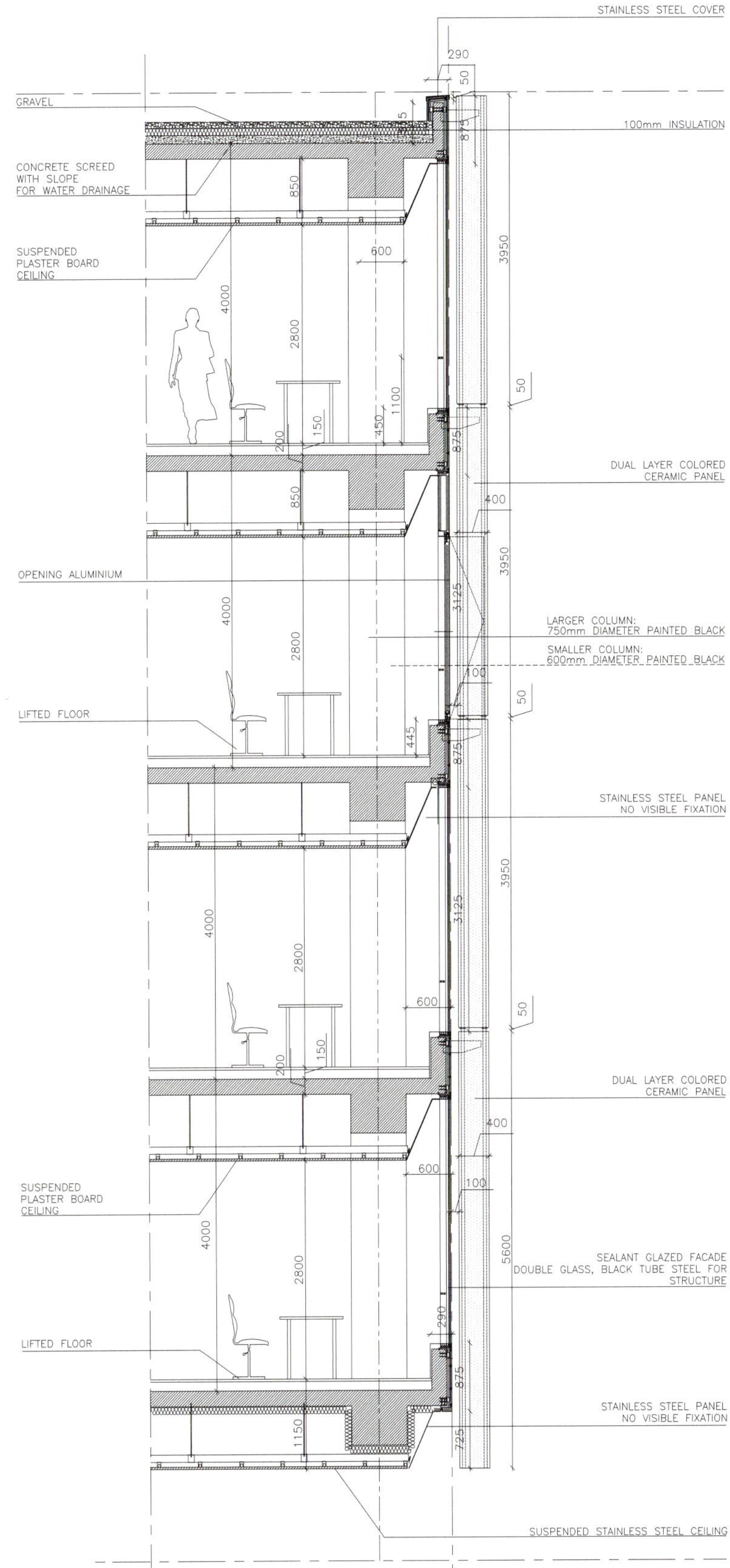

a–a' 剖面图 section a-a'

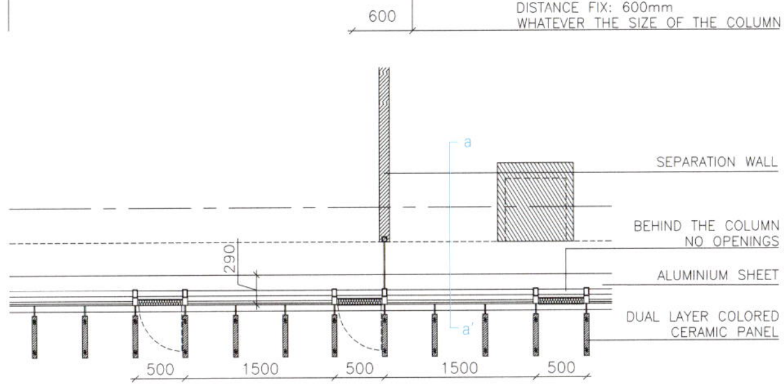

详图1 detail 1

Pevasa总部

Cuantics Creatives Lab

世界各地的组织都在面临着与扩展创新、学习和协作相关的新挑战。我们将建筑解读为这种转变的透视法，因此建筑设计应该有助于我们轻松改变日常生活和工作轨迹。因此，这个项目的根本问题是什么？未来的办公室应该是什么样的，它应该对应什么样的场所？我们的回答是参与度，即个人和集体的参与度。

我们设计了一个抱着敬畏之心适应新工作程序的内部空间，以使新的行为合乎惯例，并促成点对点模式的互动。我们的目的是超越层级分布的协作和自发组织，所以我们的关注点设定在人们的需求上。

Pevasa新总部由五个支柱支撑，并被抬升成为一个观景台，以捕获

充足的阳光并欣赏贝尔梅奥海岸边的远景和毕尔巴鄂附近令人惊叹的Ogo o海角。Pevasa总部支柱的设计代表了该公司的五个创始家族。倾斜的支柱分组形成了上部楼层和其他共享空间的入口。

支柱分布在整个平面布局中，创造出了内部阳光庭院或会议室等与众不同的用途空间。

在仙台向伊东丰雄学习了结构课程之后，我们就能够放开平面布局，以创建明亮的、有多种选择和路径的多功能空间，也有助于实现项目整体目标，鼓励协作、水平、透明和不固定用途。

建筑总共1600m^2，分两层。一层面积稍大，作为办公室；二层被设计成多功能房间，用来与Pevasa船运员工一起举办相关的会议。二楼的房间通向屋顶花园，可用于非正式聚会。建筑物的钢和混凝土基础与透明外壳形成对比，试图模糊办公室和树冠之间的界限。

Pevasa Headquarters

Organisations worldwide are facing new challenges related to scaling innovation, learning and collaboration. We understand architecture as the scenography for this transformation and hence it should be designed to ease the changes we want to bring both to our living and working routines. So the founding question in this project was? What should an office look like in the future, what premises should it respond to? Our answer was engagement, individual and collective engagement. We proposed a domestic space to respectfully accommodate the new working routines, to legitimate new behaviours and to precipitate peer to peer interaction. We aimed collaboration and self-organisation over hierarchical distributions, so our focus was set on people needs.

The new headquarters for Pevasa rests and rises over five pillars that elevate it to a lookout, capturing the light and perspectives of Bermeo shore and the amazing Ogo o cape, near Bilbao. The design of the columns supporting Pevasa headquarters alludes to each of the five founding families of this company. The pilasters are slanted and grouped to host the access to the upper floors and other shared uses. The columns distribute the whole floor plan and they create distinctive uses as interior light courtyards or meeting rooms. The structural lesson learned from Toyo Ito at Sendai allowed

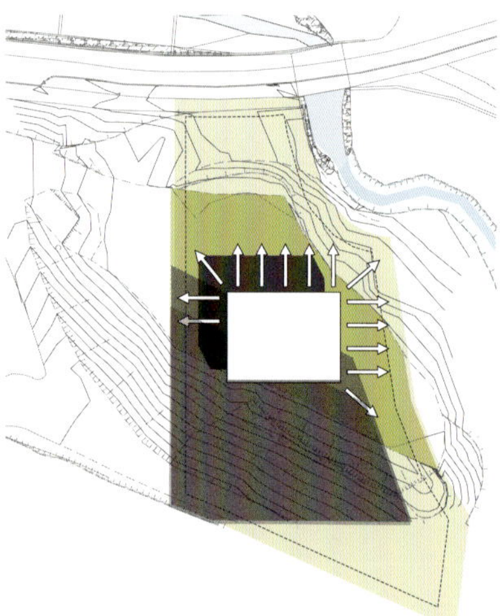

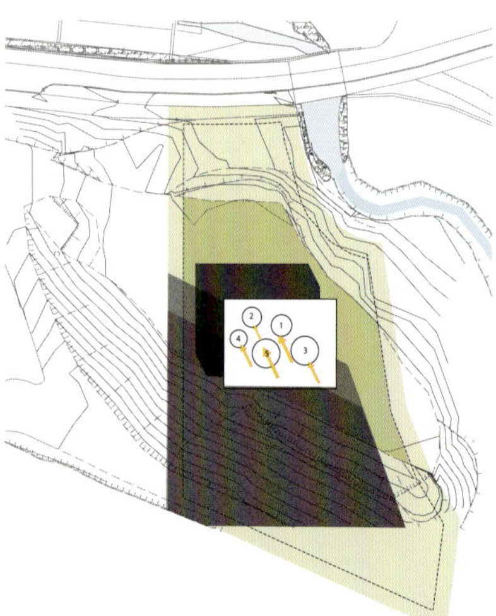

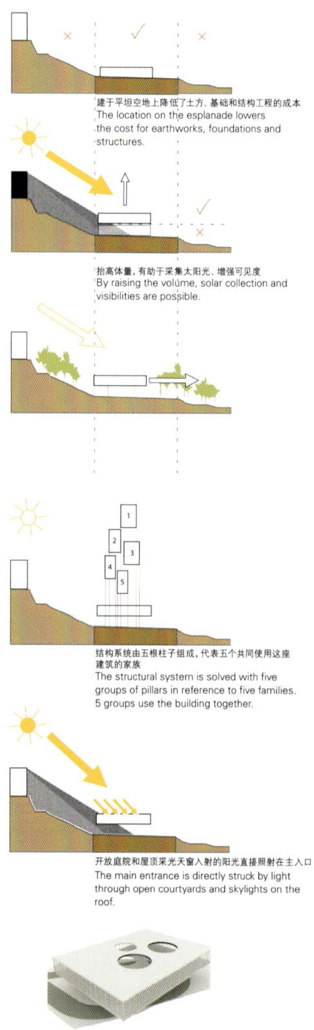

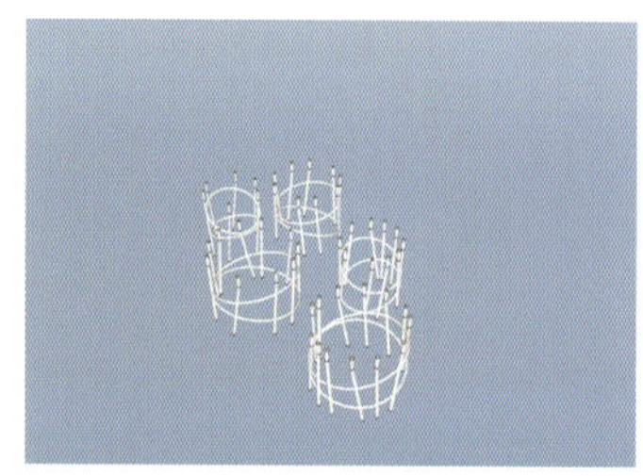

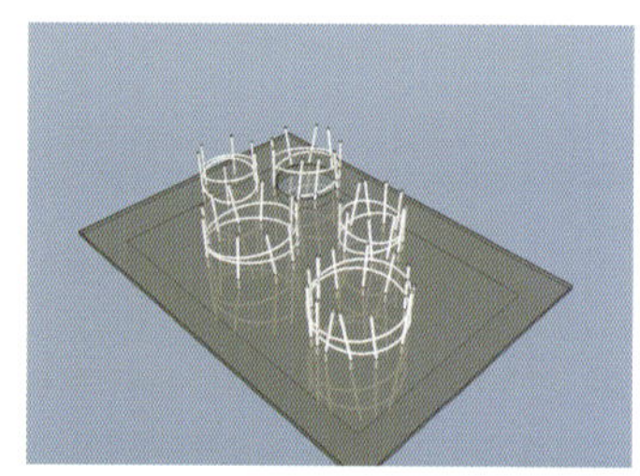

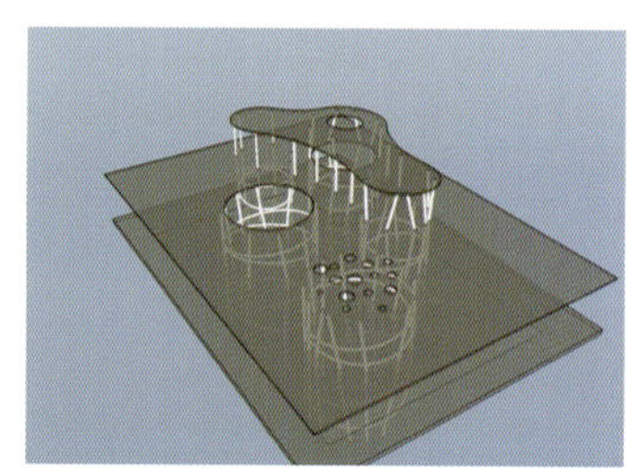

us to free the floor plan to create bright and versatile spaces with multiple options and paths, that contribute to the overall purpose to encourage collaborative, horizontal, transparent and fluid uses.

A total of 1,600 square meters are distributed on two floors. A larger first floor for offices, and a second floor occupied by a multipurpose room designed to host conferences or meetings with the crews of Pevasa ships. The upper room opens to a roof garden for informal encounters. Steel and concrete foundation contrasts with the transparent outer shell of the building, conceived to blur the limits between the office and the tree canopy. Diego Soroa

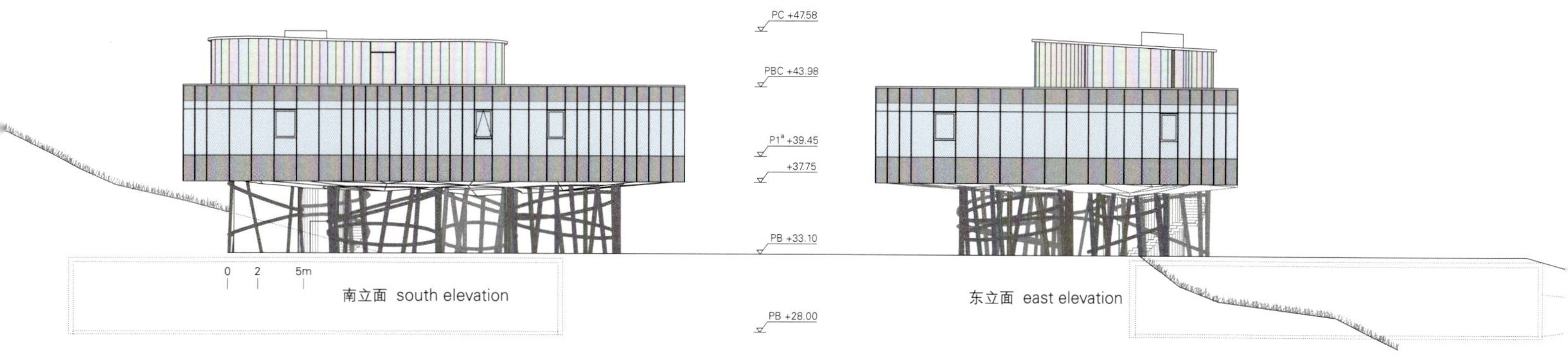

南立面 south elevation

东立面 east elevation

A–A' 剖面图 section A-A'

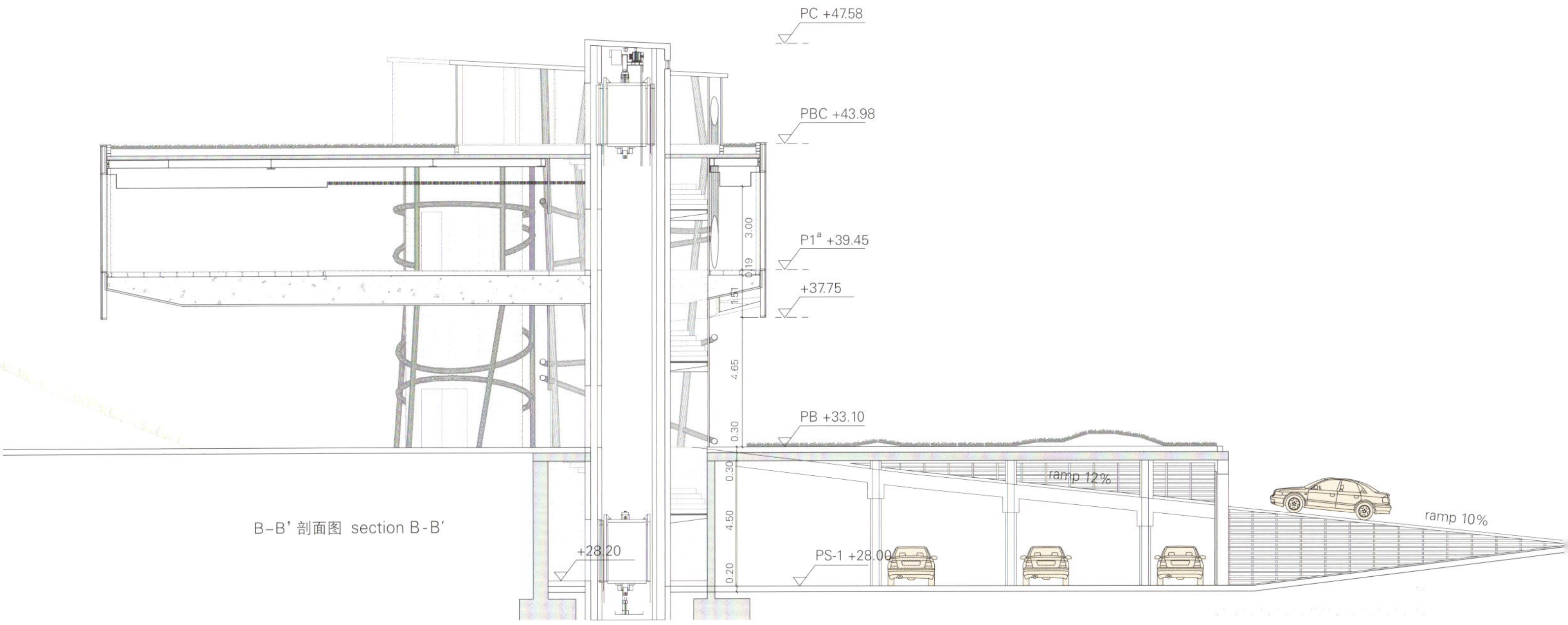

B-B' 剖面图 section B-B'

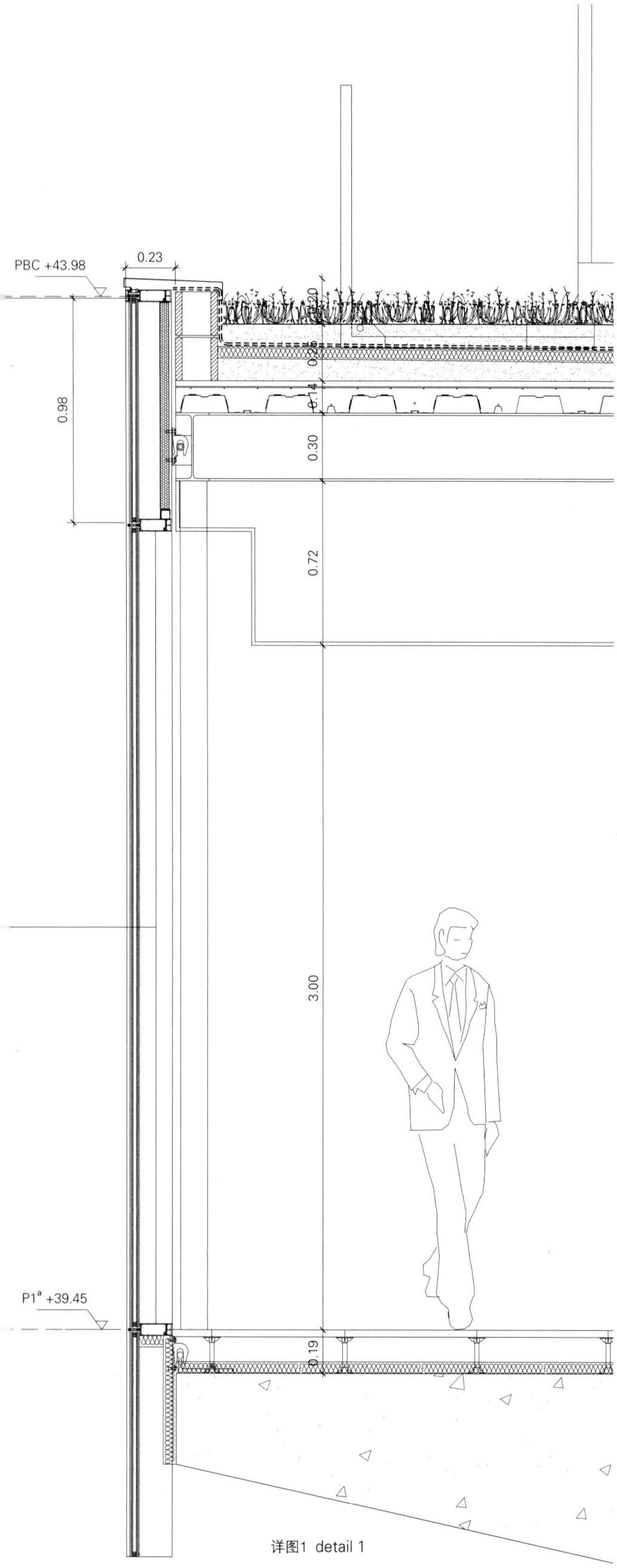

详图1 detail 1

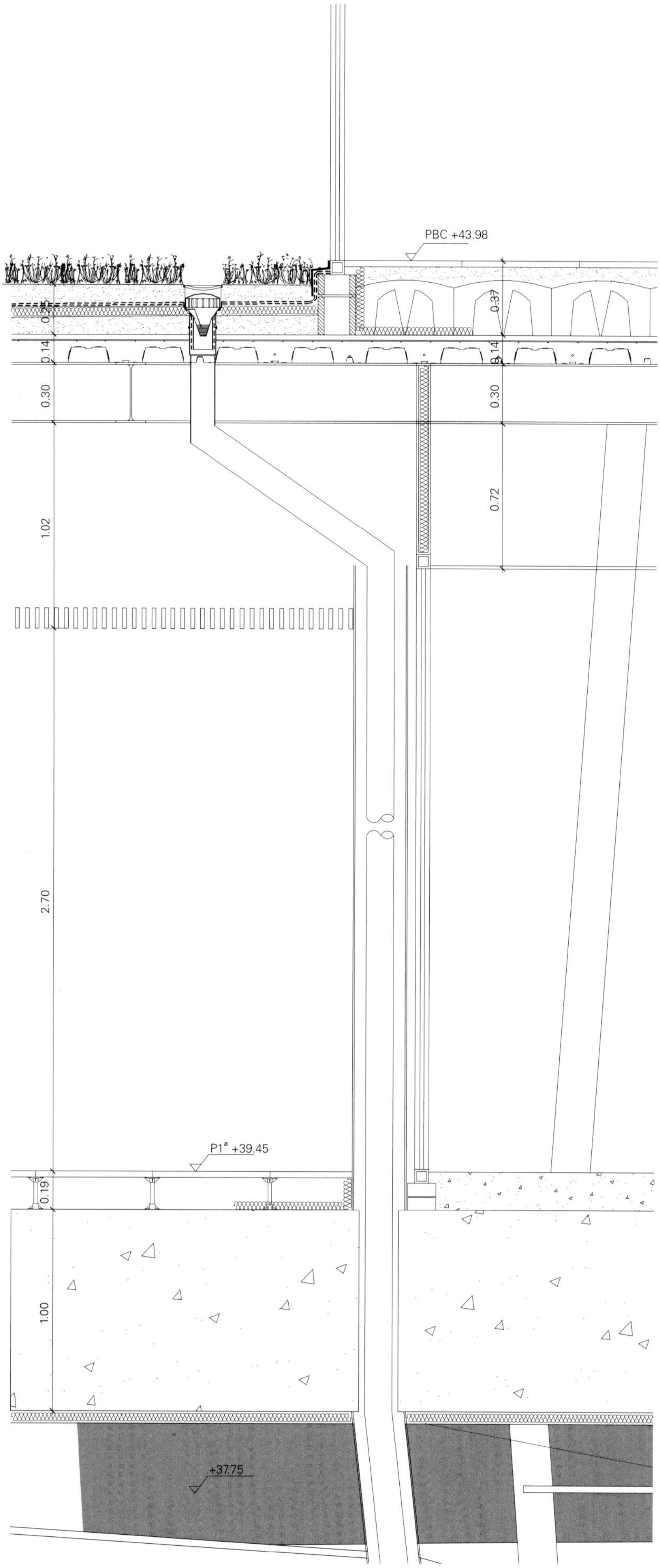

详图2 detail 2

SALIDA

项目名称：Pevasa Headquarters
地点：Bermeo, Bizkaia, Spain
建筑师：Cuantics Creatives
总建筑师：Diego Soroa
工程师：LKS Ingeniera - Mikel Txintxurreta, Igone Asua y Patxi Larozgoitia
施工：Ferrovial / 客户：Pevasa
用地面积：3,757m² / 建筑面积：814m² / 总建筑面积：1,679m²
设计时间：2013 / 竣工时间：2015
摄影师：©Wenzel (courtesy of the architect)

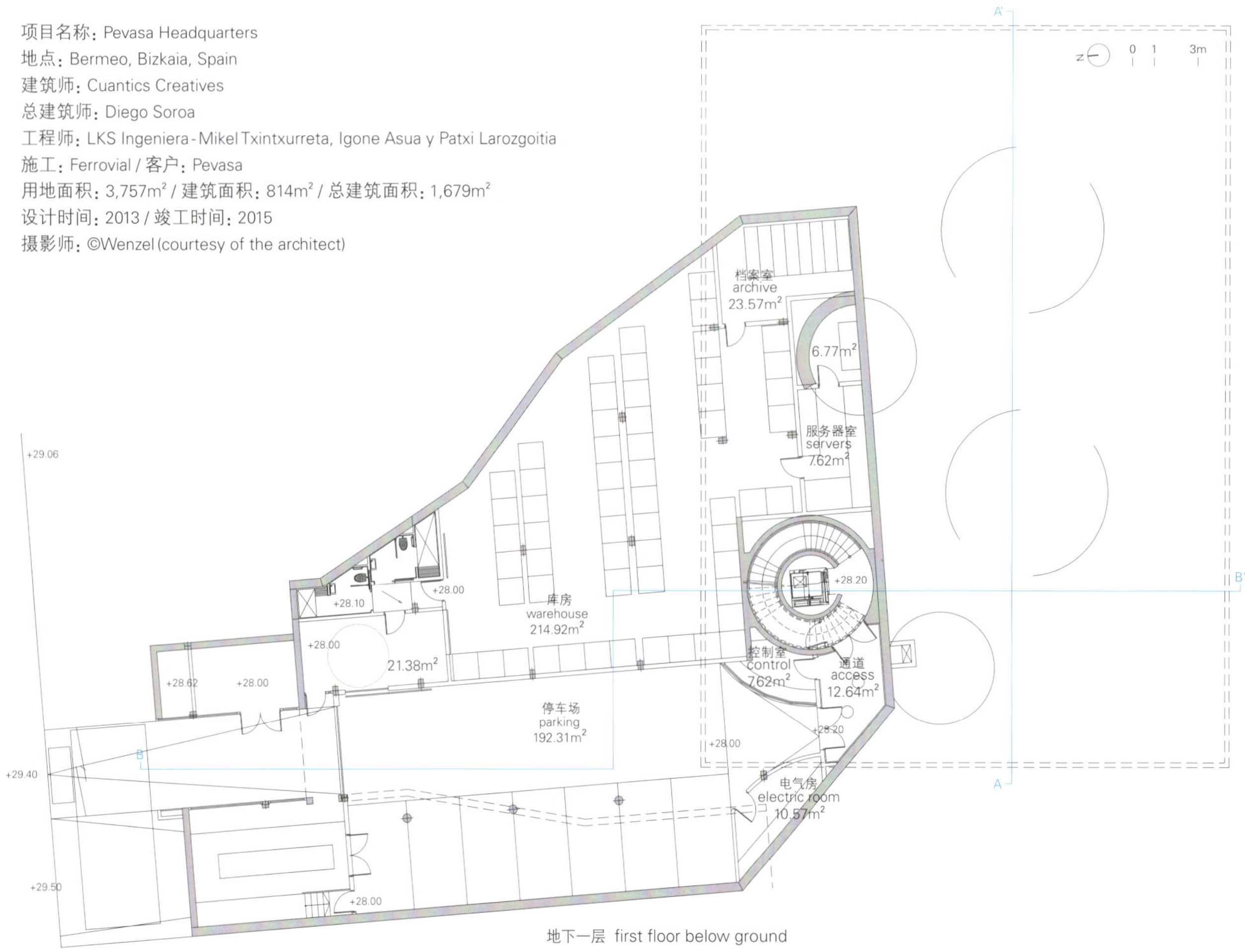

地下一层 first floor below ground

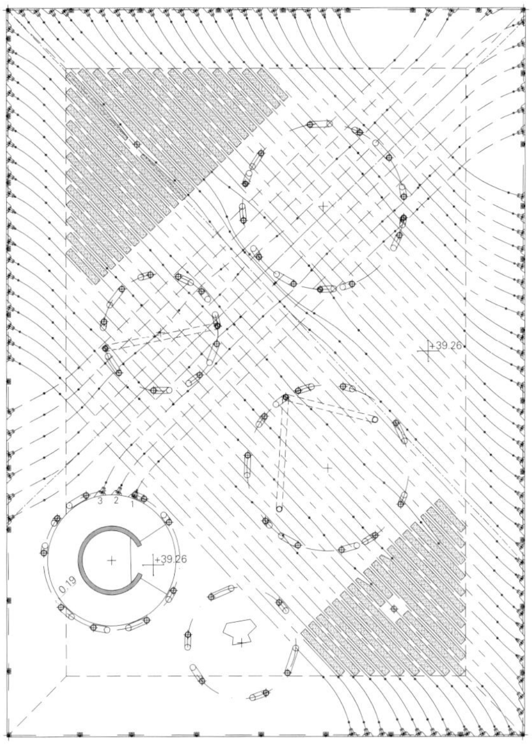

楼层详细结构 detail structure of the floor

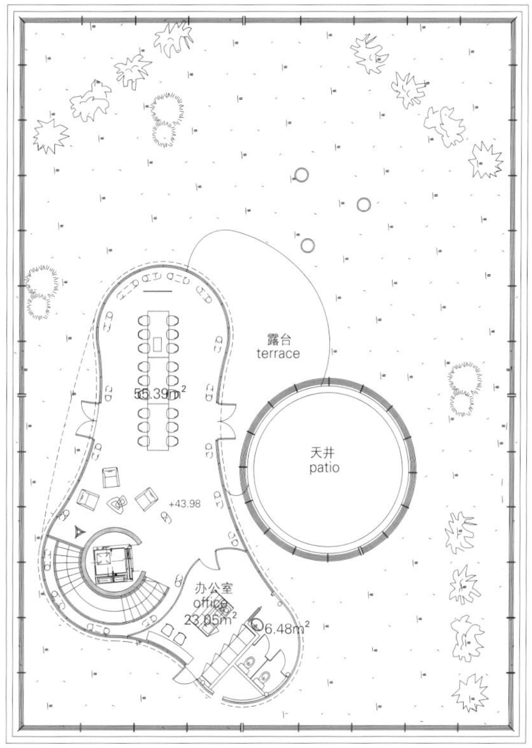

三层 third floor

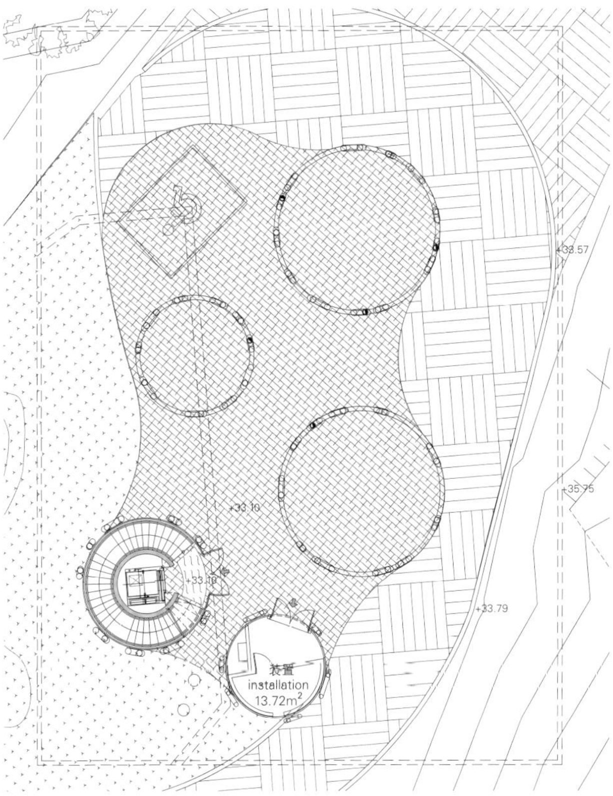

一层 first floor

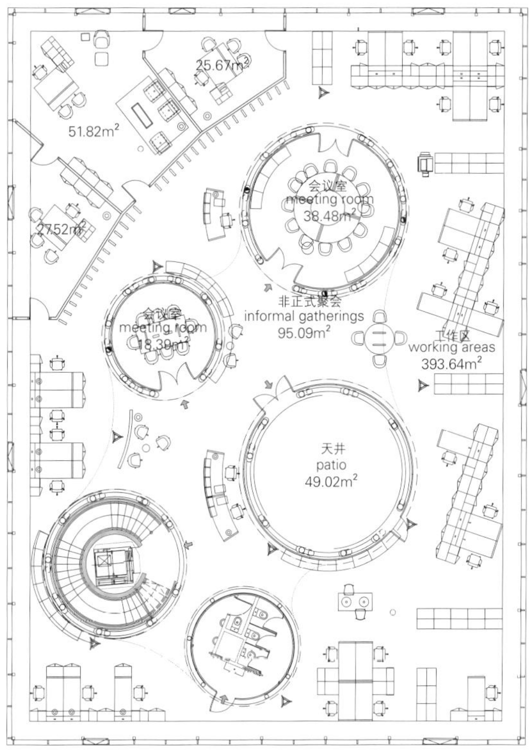

二层 second floor

泰国国家石油勘探生产公司S1办公楼

Office AT

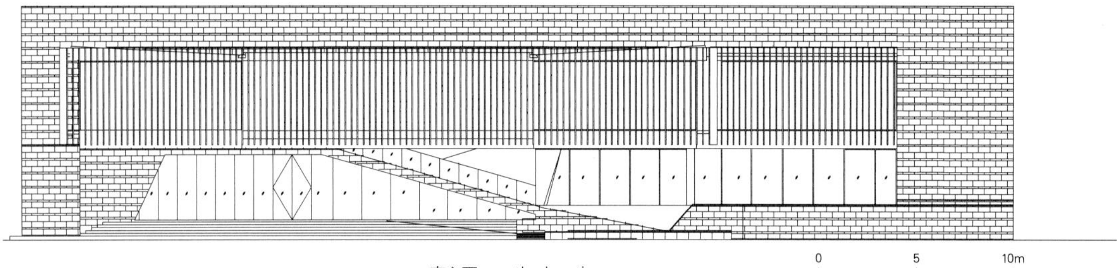

南立面 south elevation

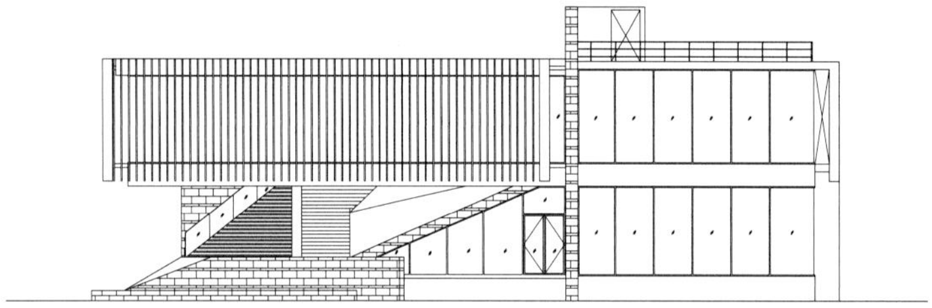

东立面 east elevation

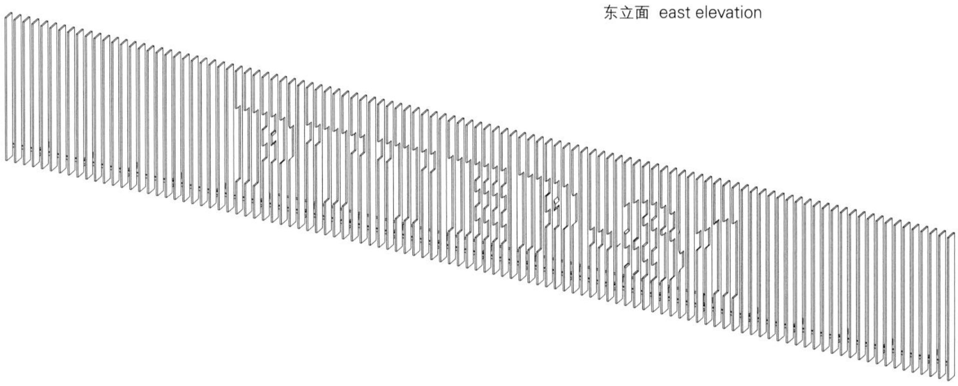

PTTEP是泰国的国有石油勘探和生产公司，致力于为泰国提供可持续的石油供应资源。其位于泰国最大的陆地油田，油田以诗丽吉王后的名字命名，位于泰国北部甘亰碧省东北方向的Lankrabue。此地已被持续开采运营了30年。2011年，PTTEP举办了一场建筑设计竞赛，要求参赛者设计一幢两层的办公楼，来纪念这段长达30年的石油勘探历程。最终，Office AT成了这次设计竞赛的赢家。

场地位于诗丽吉油田的入口处，被一批小型建筑包围。PTTEP公司希望建造一栋占地2000m²的两层办公大楼作为该油区的主要办公楼。地面是此设计的一个重要元素。这是对地底石油的一种象征表现。通过将地面层抬升，由上覆草皮的折叠混凝土板覆盖的地面下方的空间，可作为建筑入口。

除了将折叠混凝土屋顶板下方的一层作为建筑的入口之外，这里还包含了其他功能部分，例如，教室、接待区等。新的折线形地面层的顶部空间功能灵活。平日里，它可以用作楼梯；在活动期间也可变为表演场地。

一层如之前所述，在设计中是地面的象征。而二层则表现了PTTEP公司的创新性。当这样的二层位于一层之上时，造就了办公室与地面的“碰撞”。

在将二层的办公室安置在折叠混凝土板的顶部后，为使地面层仍能与二层办公室连接，设计师将办公区从中间分开，空出一个中心庭院的空间。抬升的地面层与二层办公室可以直接接触。这样一来，办公室的员工就可以出去享受户外空间，并饱览庭院景致。在典型的办公建筑中，还从未出现过将二层办公空间直接与底层相连的尝试。PTTEP新大楼将改变这一传统认知。

在二层，也有一座连接两部分办公室的廊桥。建筑中的办公区域都覆盖着铝质散热片，可以为建筑遮光并提高其节能性能。

PTTEP-S1 Office

PTT Exploration and Production Public Company Limited (PTTEP) is a Thailand national petroleum exploration and production company dedicating to provide a sustainable petroleum supply to Thailand. The biggest petroleum supply on land, called Sirikit oil field is located on Lankrabue , the northeastern district of Kamphaeng Phet Province, northern of Thailand. The site is operated for 30 years. In year 2011 PTTEP started a competition to design a 2-storey office building to commemorate 30 years of petroleum exploration and Office AT was a winner for the design competition.
The site is located at the entrance of Sirikit oil field surrounding with small buildings. The PTTEP wants to build a 2000 s.q.m. 2-storey office building as a main office for the site.
The ground is an important element of the design. It is a representation of "Oil" underground. By shifting the ground up, the space under the ground plane which is made of folded

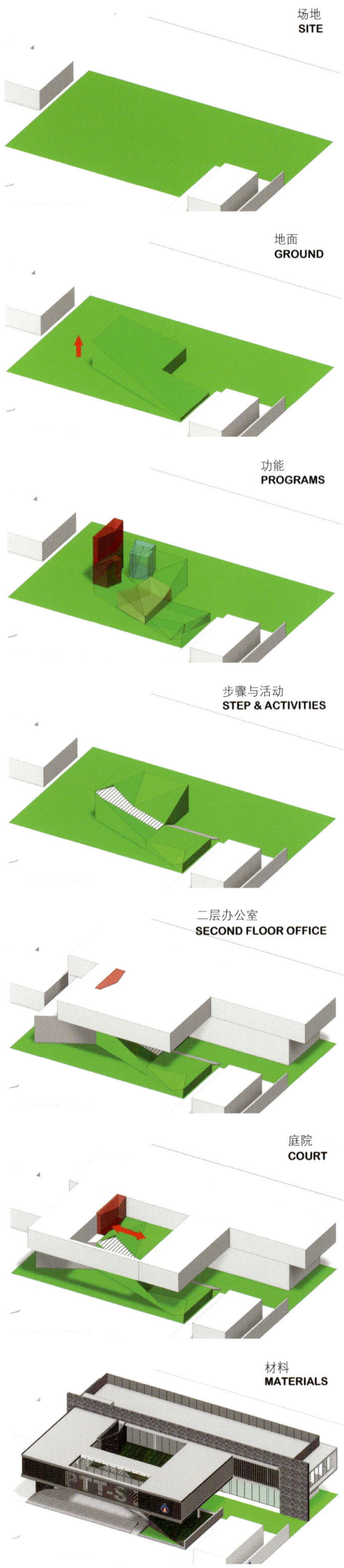
场地
SITE
地面
GROUND
功能
PROGRAMS
步骤与活动
STEP & ACTIVITIES
二层办公室
SECOND FLOOR OFFICE
庭院
COURT
材料
MATERIALS

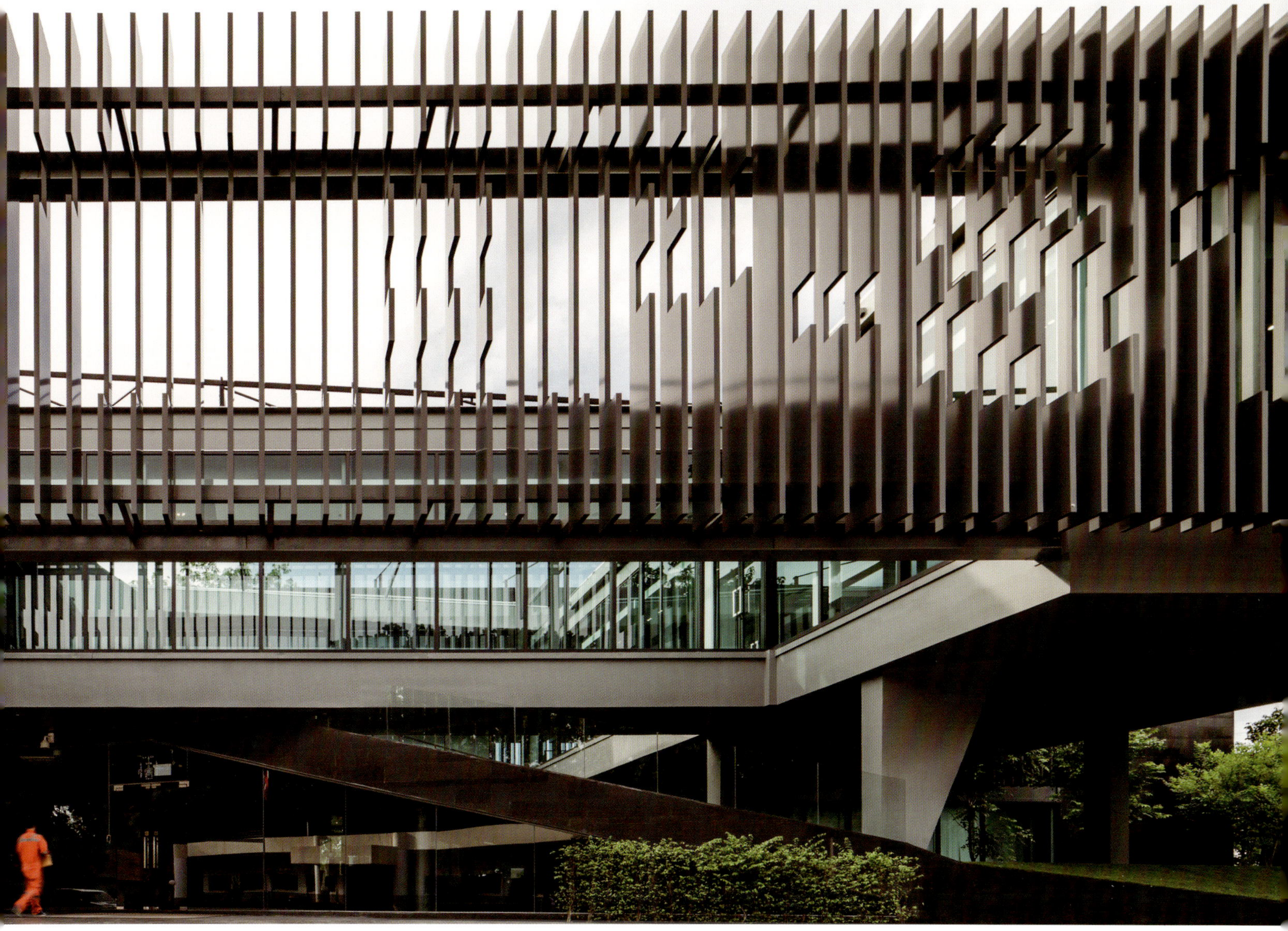

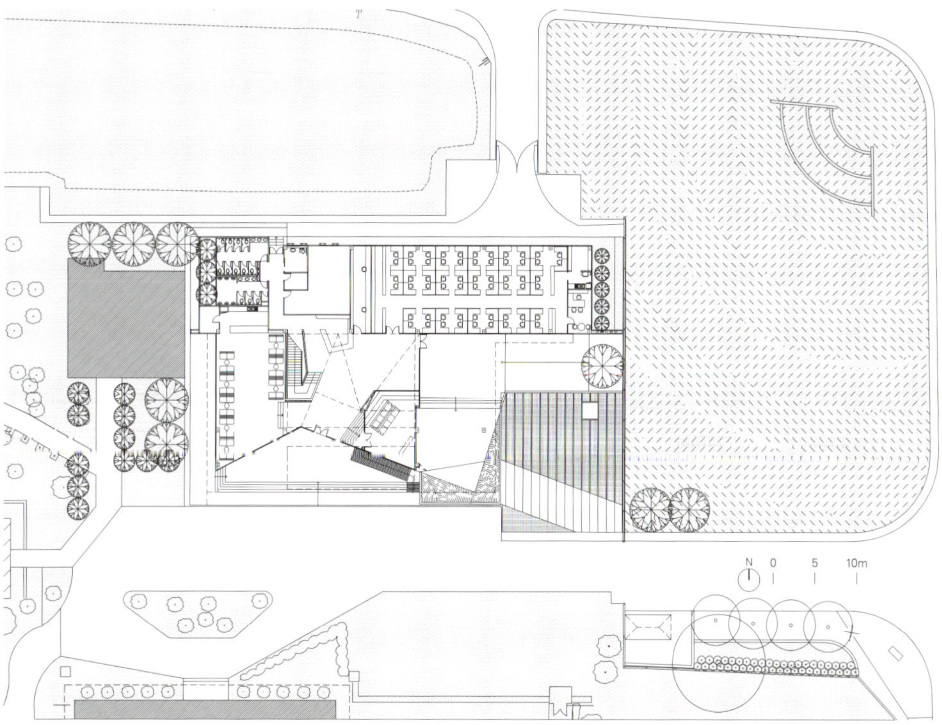
N
0
5
10m

concrete slab and covered with grass can be used as an entrance to the building.

The program is not only the entrance for the building, on the ground floor under the folded concrete roof slab, it also contains with other programs such as lecture room, reception area, etc.

The top of the new folded ground plane also can be used as the flexible space. It can be used as the stair in normal times and can be turned into the performance space during the event.

The first floor, as mentioned before is a represent for the ground. For the 2nd floor, it is representing PTTEP as an innovative company. When the 2nd floor is put on top of the 1st floor, finally they are meeting each other between "office and ground".

In order to connect the ground plane with office space on the

2nd floor, the office is divided at the center to have a courtyard. The ground plane that is shifted up can directly touch the 2nd floor office. Office users can go outside to enjoy the outdoor space and can have a view of the courtyard. For the typical building, the office space on the second floor never has the direct relationship with the ground floor. The new building for PTTEP will change this perception.

On the 2nd floor, there is also a bridge connecting 2 parts of the office. Office areas in the building are covered with aluminium fins to protect the building from sun light and increase energy saving performance.

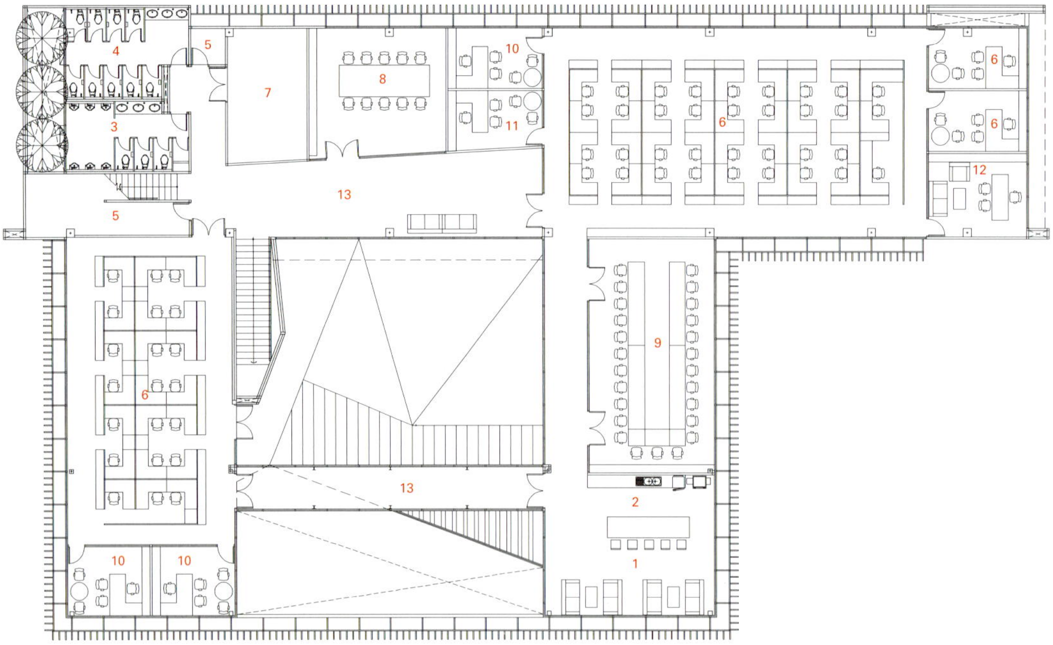

1. 休息厅 2. 食品室 3. 男卫生间 4. 女卫生间 5. 储藏室 6. 办公室 7. IT基础设施 8. 会议室 9. 大会议室
10. 主管办公室 11. 高级职员办公室 12. 副主席办公室 13. 走廊

1. lounge 2. pantry 3. men wc 4. women wc 5. storage 6. office 7. IT infrastructure room 8. meeting room
9. large meeting room 10. superintendent's room 11. senior stuff's room 12. vice president's room 13. walkway

二层 first floor

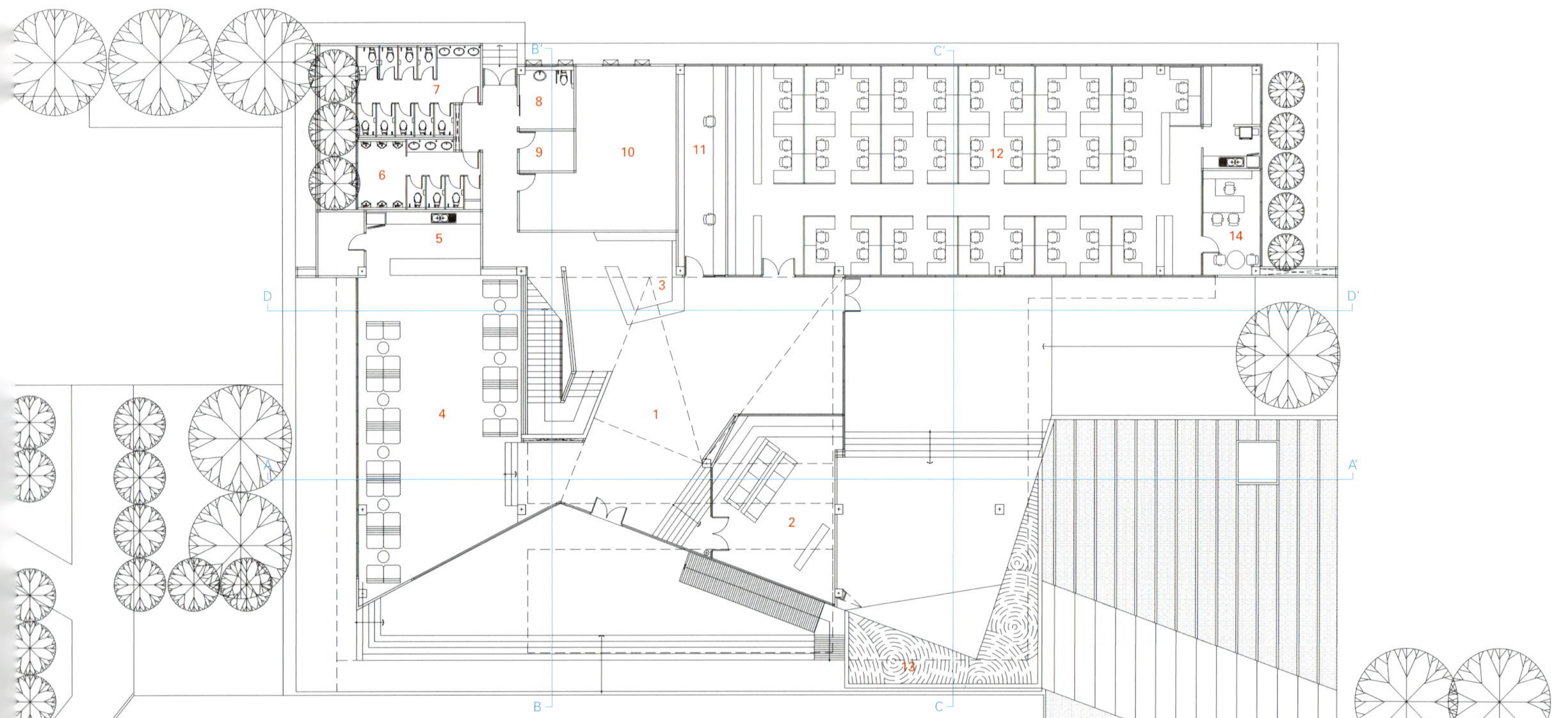

1. 门厅 2. 入职安全教育室 3. 信息处 4. 休息厅 5. 食品室 6. 男卫生间 7. 女卫生间 8. 残障人士卫生间
9. 电气室 10. 储藏室 11. 信息交流处 12. 办公室 13. 水池 14. 主管办公室

1. lobby 2. safety induction area 3. information 4. lounge 5. pantry 6. men wc 7. women wc 8. disabled wc
9. electrical room 10. storage 11. info & communication 12. office 13. pool 14. superintendent's room

一层 ground floor

EXIT

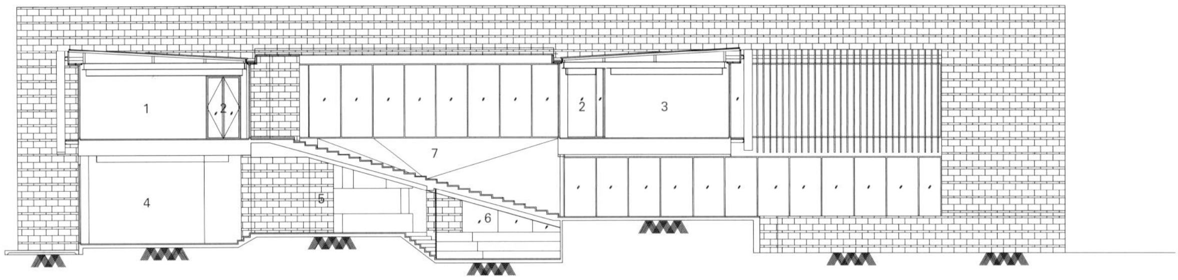

1. 办公室 2. 走廊 3. 会议室 4. 休息厅 5. 门厅 6. 入职安全教育室 7. 庭院

1. office 2. walkway 3. meeting room 4. lounge 5. lobby 6. safety induction area 7. court

A–A' 剖面图 section A-A'

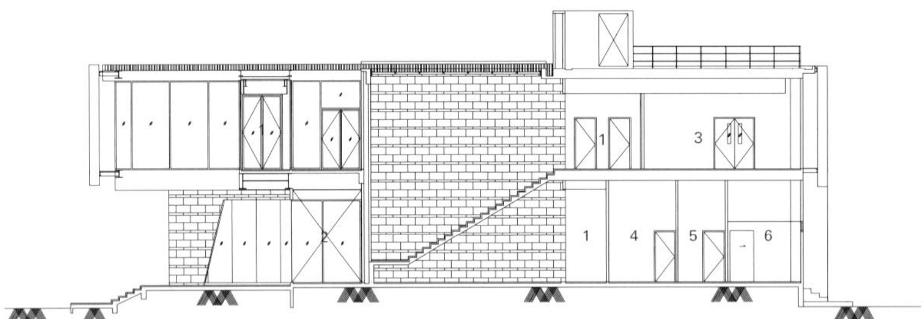

1. 走廊 2. 门厅 3. IT基础设施 4. 储藏室 5. 电气室 6. 残障人士卫生间

1. walkway 2. lobby 3. IT infrastructure 4. storage 5. electrical room 6. disabled wc

B–B' 剖面图 section B-B'

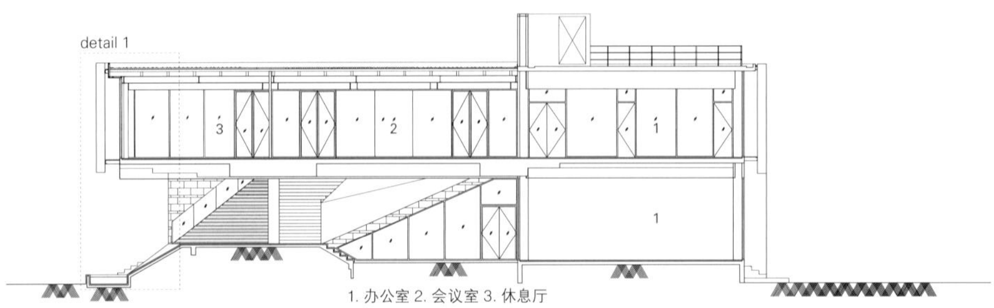

1. 办公室 2. 会议室 3. 休息厅

1. office 2. meeting room 3. lounge

C–C' 剖面图 section C-C'

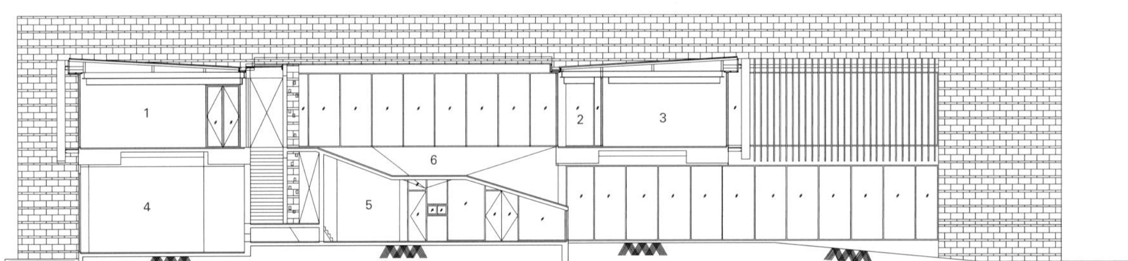

1. 办公室 2. 走廊 3. 会议室 4. 休息厅 5. 门厅 6. 庭院

1. office 2. walkway 3. meeting room 4. lounge 5. lobby 6. court

D–D' 剖面图 section D-D'

项目名称：PTTEP-S1 office / 地点：Lankrabue, Kampaengphet, Thailand / 建筑师：OFFICE AT Co., Ltd. / 项目建筑师：Surachai Akekapobyotin / Juthathip Techachumreon / 建筑团队：Pinyo Tanadirek / 室内设计师：OFFICE AT Co., Ltd. / 结构工程师：Mr. Sarawut Yuanteng / 系统工程师：COU.Co., Ltd. / 承包商：Italian Thai Engineer / 客户：PTTEP Siam Limited / 楼层：2floors / 面积：2,000m² / 施工造价：THB 60million / 施工时间：2011—2014 / 材料：transparent glass, aluminium fin, stone / 摄影师：©W-Workspace (courtesy of the architect)

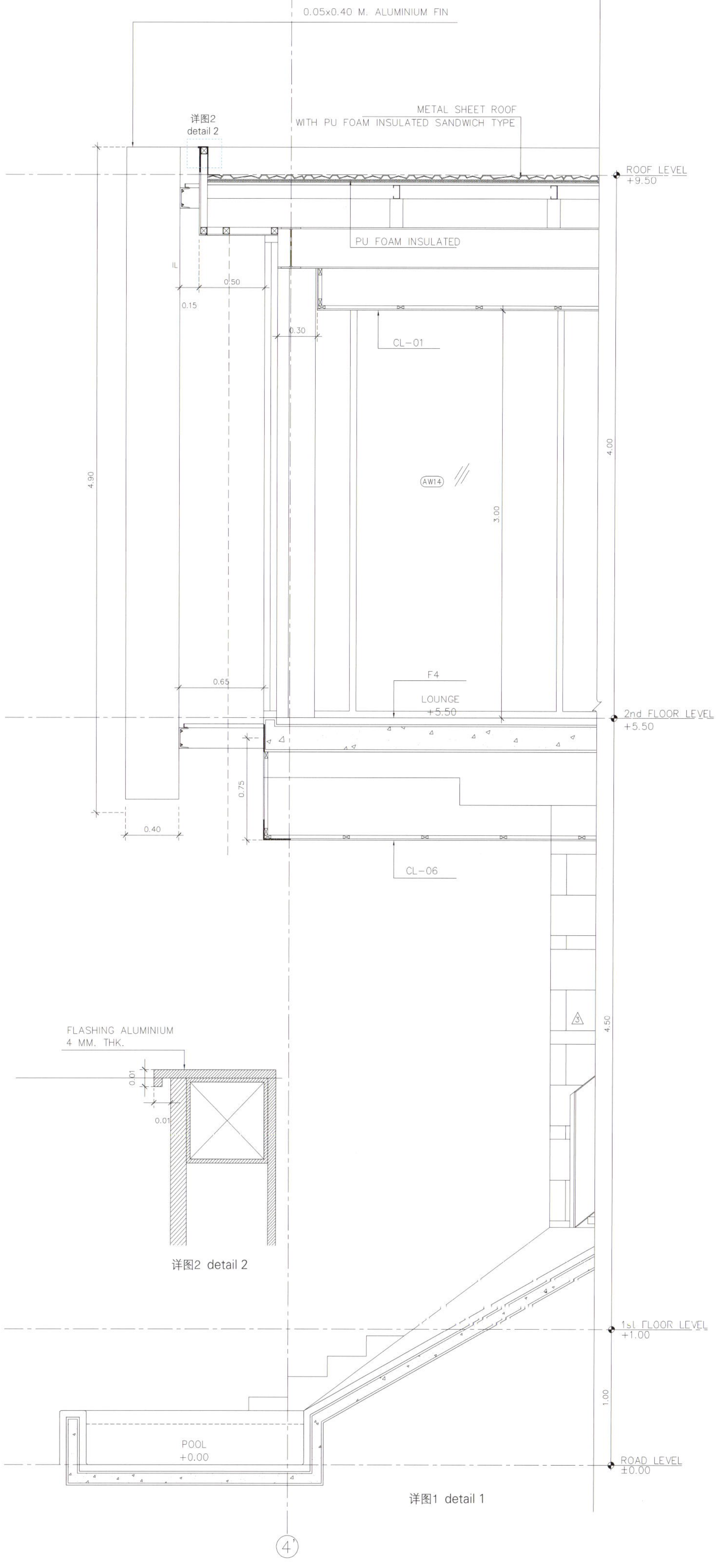

详图2 detail 2

详图1 detail 1

杜嘉班纳公司办公楼

Piuarch

该地段的狭长形状和既有建筑物的体量，形成了包括三座平行建筑物的组合方案。皮亚韦大街上的建筑及其后方建筑的地上部分都是七层高。中间的建筑通过一系列楼梯和人行天桥与其他建筑连接在一起，它有三层楼，悬浮在一个连接于其他三座建筑的体量的上方，这个体量垂直于皮亚韦大街，在地面上有四层高。通过这种方式，建筑师创建了一个宽敞的大进深的内部庭院，这个庭院可以从所有建筑物俯瞰到。该项目还设计了地下三层，其中两层用于停车场。

建筑物的复杂结构反映了空间的多种不同的用途，既有办公室，也有客户接待区域和娱乐区域，有用于不同楼层举办时装表演的准备协调的区域，一楼还有可以通向庭院的公共区域和会议室。

皮亚韦大街一侧的立面采用了3.60m×3.60m的方形百叶窗模块系统，以交错重叠的方式布置，由不锈钢板条装饰，和邻近的大都会酒店的立面相似。在夜间，外立面在室内灯光的映衬下充满了生气，窗户也显得五彩斑斓，透明玻璃与灰色调的玻璃相互交替，与那些打开的和固定的窗户互为呼应。这些立面有众多开窗的建筑坐落于基柱上，基柱的大小和材料与附近的大都会酒店极为相似，营造出一种强烈的视觉连续性。

这样的风格也用在了临街的立面上，玻璃和金属的使用传达出透明和轻盈的感觉，与室内立面的简朴风格形成对比。皮亚韦大街一侧立面的方形模块的主题在这里得到了重复使用，无论在尺寸还是形状上都是重复的，不过在这里是由玻璃纤维混凝土（GRC）制成的，并安装了玻璃门窗。

办公室和连接空间的室内设计特点是采用了简单而实用的布局，布置成开放空间。水泥天花板按照木板条的形状加工成形，采取裸露在外的形式铺设，并安装了配有空调和照明系统的横梁。地面材料多种多样：在办公室采用了陶瓷瓷砖，时装秀协调区域采用了橡木地板，客户接待和娱乐区域则是用回收的旧木材制作的。在一层，无论是在室内还是庭院，各处都由一种黑火山岩石：玄武岩铺设。在这里，就像为杜嘉班纳公司完成的其他项目一样，玄武岩的采用表达了鲜明的地中海感觉，成了这家时装公司的招牌风格。

Dolce & Gabbana Office

The long, narrow shape of the lot and the volume of the pre-existing building have led to a compositional solution involving three parallel buildings. The building on Viale Piave and the one farther back rise seven floors above ground. The central body, connected to the rest of the construction by means of a system of stairs and skyways, has three floors and is suspended over a body that connects the three buildings, perpendicular to Viale Piave and rising four floors above ground. This way a large, deep inner courtyard is created, overlooked by all the buildings. The plan also provides for three underground floors, two of which will be used for parking.

The complex structure of the building reflects the variety of uses of the spaces, with offices, areas for receiving guests and entertainment and coordination areas for the preparation of fashion shows located on various floors, and with common areas and meeting rooms on the ground floor, opening onto the courtyard.

The facade on Viale Piave has a system of brise-soleil square modules, 3.60x3.60m, arranged in a staggered, overlapping

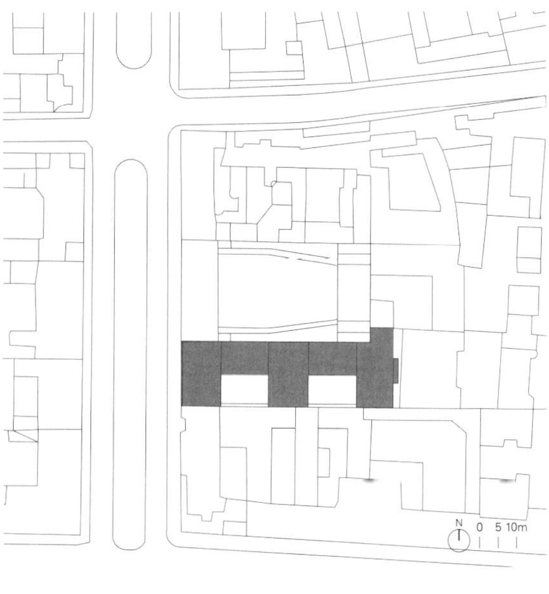

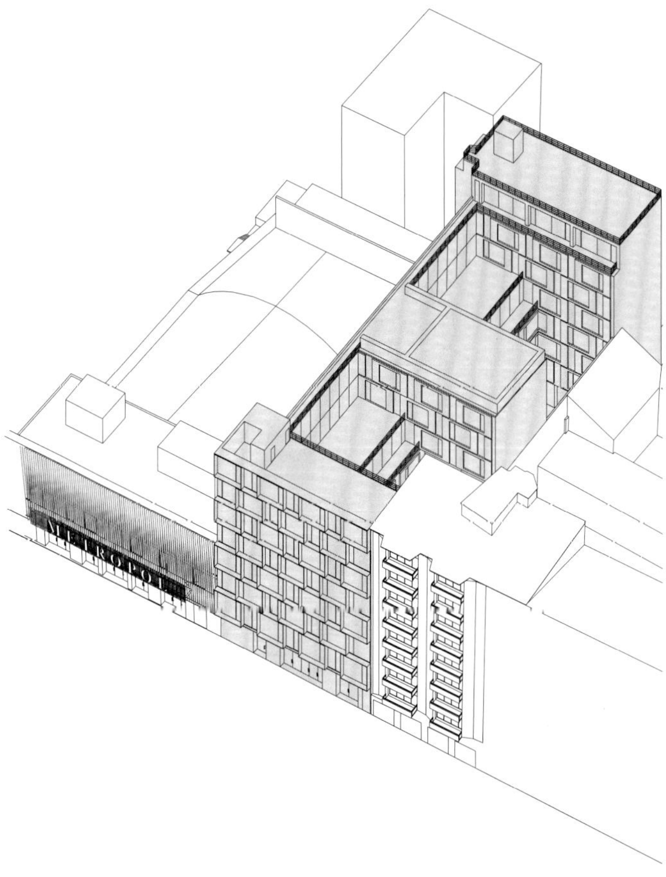

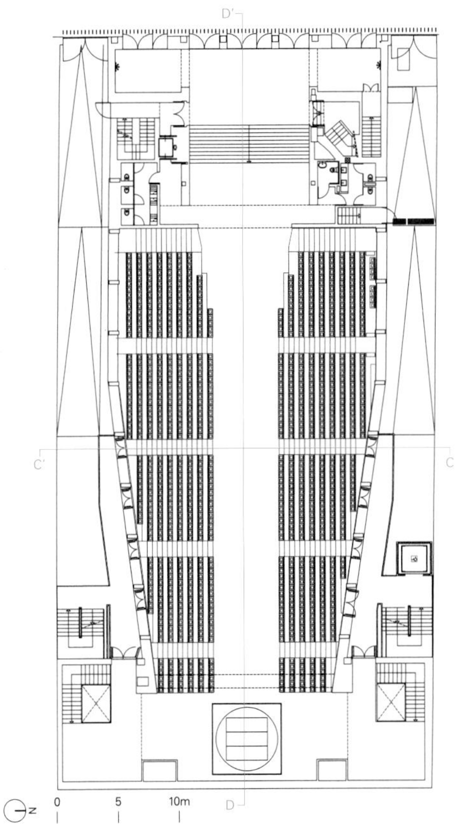

一层——大都会酒店
ground floor_metropol

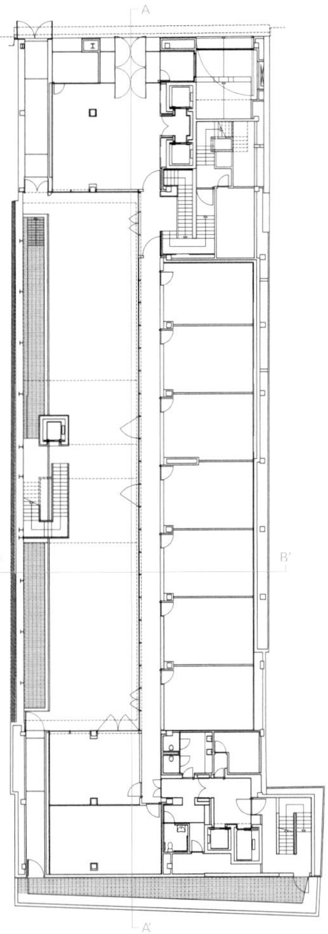

一层——办公室
ground floor_office

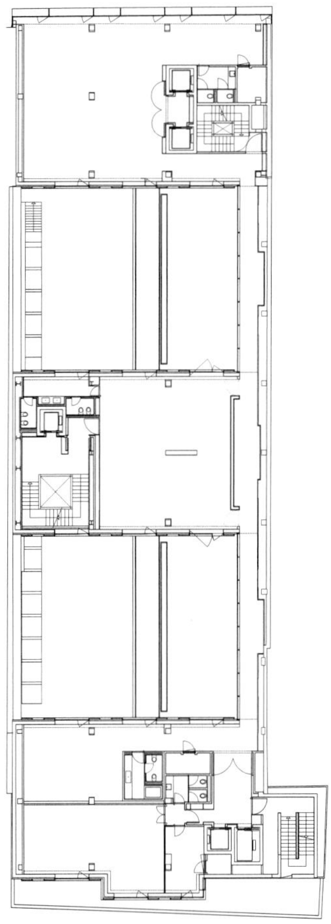

五层——办公室
fourth floor_office

manner, made from stainless steel like the slats on the facade of the adjacent Metropol space. In the nighttime the interior lighting animates the facade and brings out the chromatic pattern of the windows, in which transparent glass alternates with glass in gray tones, corresponding to windows that open and those that are fixed. The seven floors with windows stand on a socle that is similar in size and materials to that of the Metropol, establishing a strong visual continuity.

The style used for the front on the street, which conveys a sense of transparency and lightness through the use of glass and metal, contrasts with the more austere style of the interior facades. Here the theme of square modules of the facade on Viale Piave is repeated, in size and geometry, but made from Glass Reinforced Concrete(GRC), with glass doors and windows.

The interiors of the offices and connecting spaces are characterized by a simple and functional layout, organized into open spaces. The cement ceilings, molded with wooden slats, are left exposed, and have beams equipped with HVAC and lighting systems. A variety of materials are used for the floors: stoneware tiles in the offices, oak floorboards in the fashion show coordination areas and old reclaimed woods in the areas for receiving guests and entertainment. On the ground floor, both in the interior and in the courtyard, basaltina, a dark volcanic stone, is laid everywhere. Here, as in other projects done for Dolce&Gabbana, the aesthetics of the basaltina expresses a definite Mediterranean feeling, which becomes a strong reference for the fashion house.

A–A' 剖面图 section A-A'

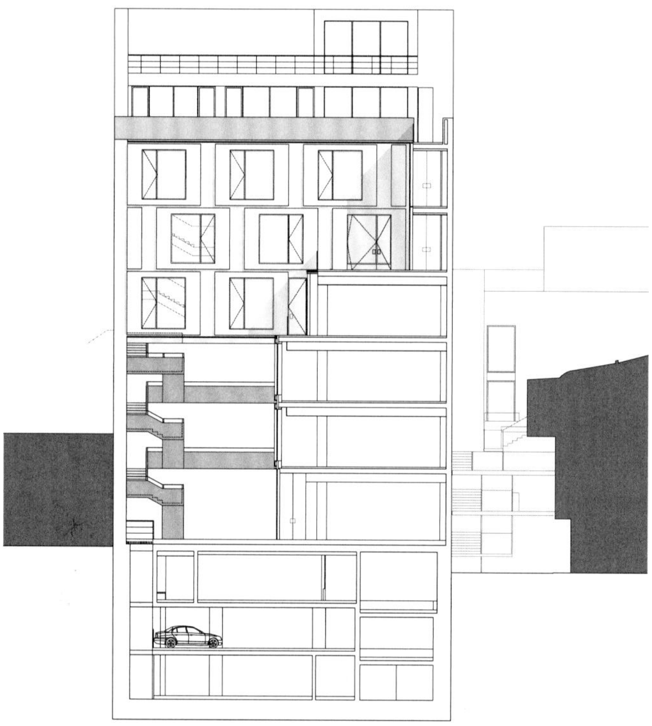

B–B' 剖面图 section B-B'

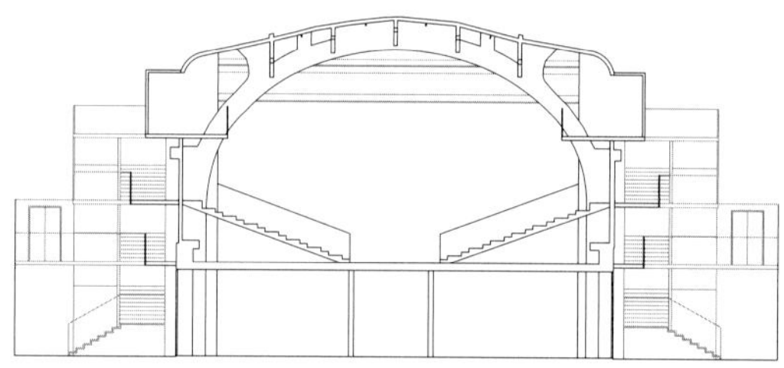

C–C' 剖面图 section C-C'

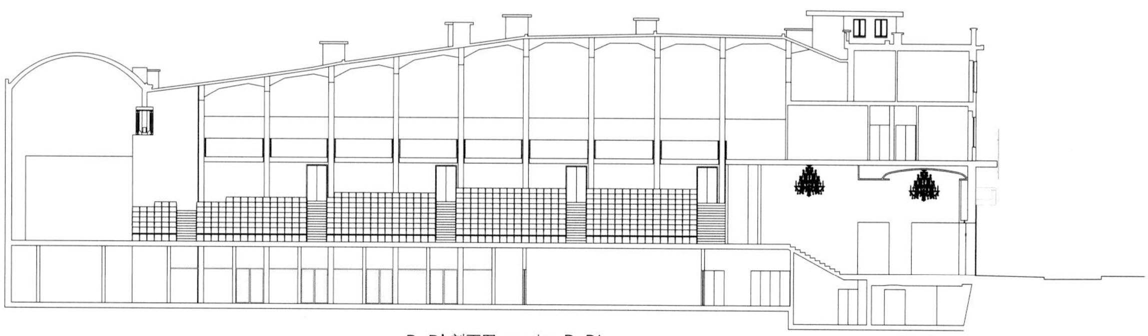

D–D' 剖面图 section D-D'

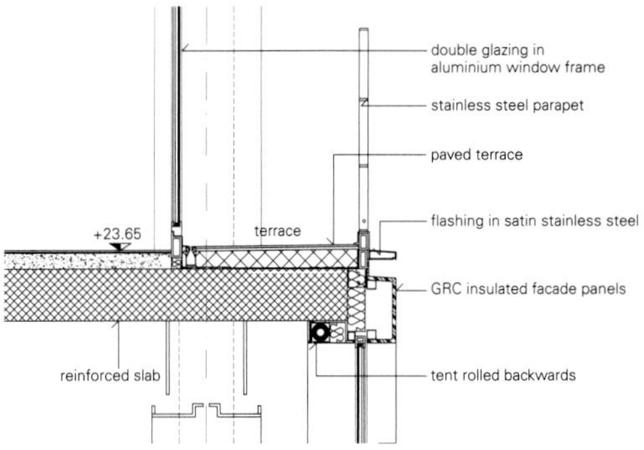

详图1 detail 1

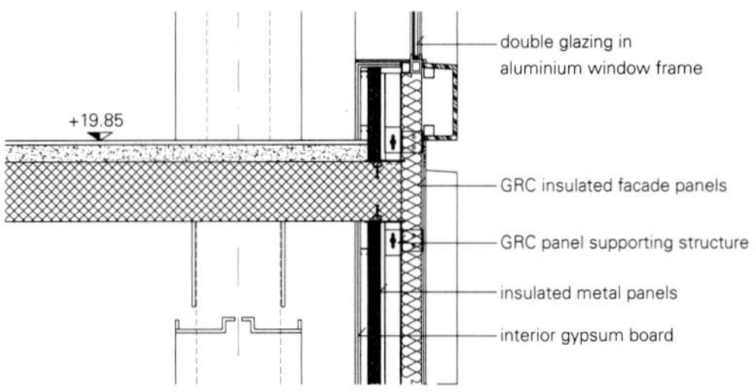

详图2 detail 2

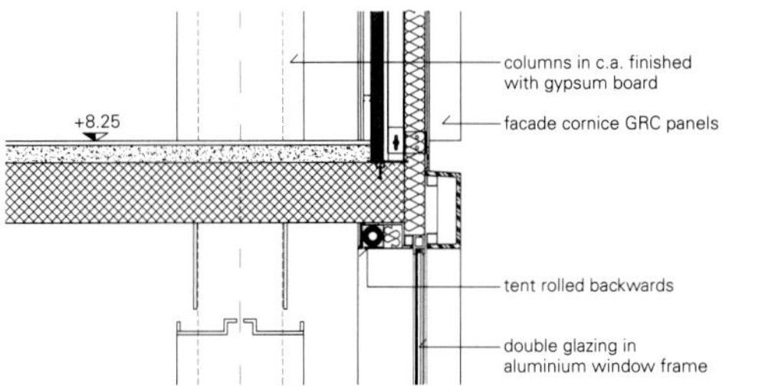

详图3 detail 3

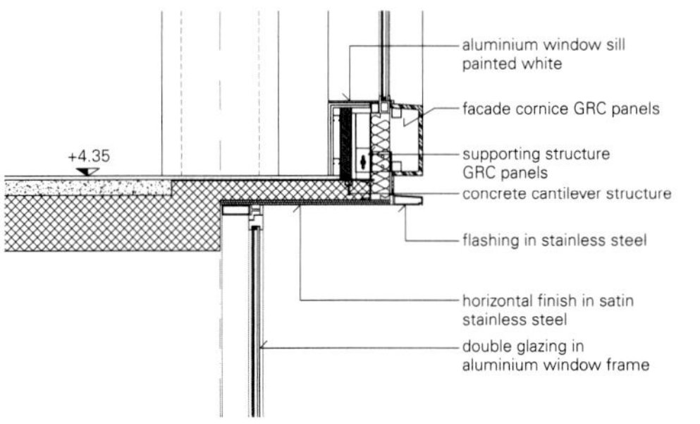

详图4 detail 4

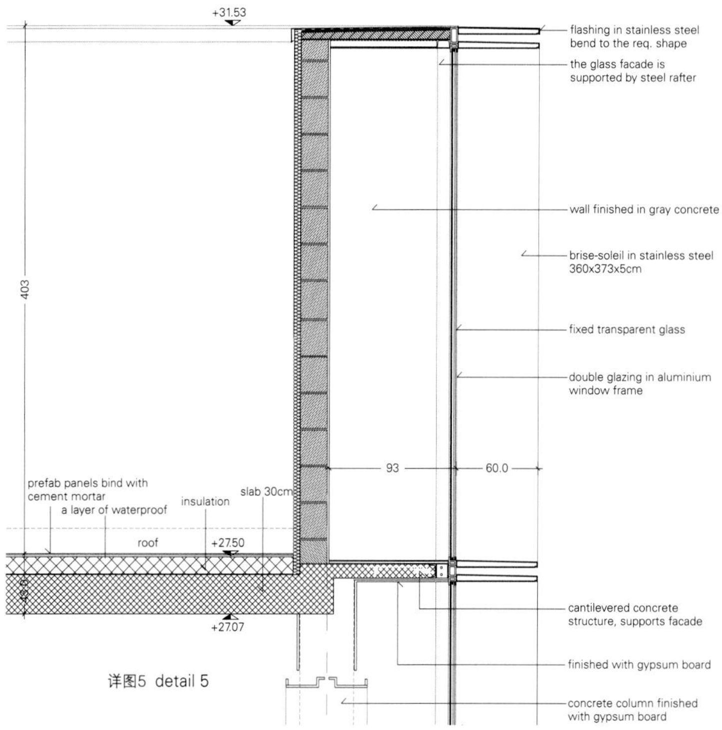

详图5 detail 5

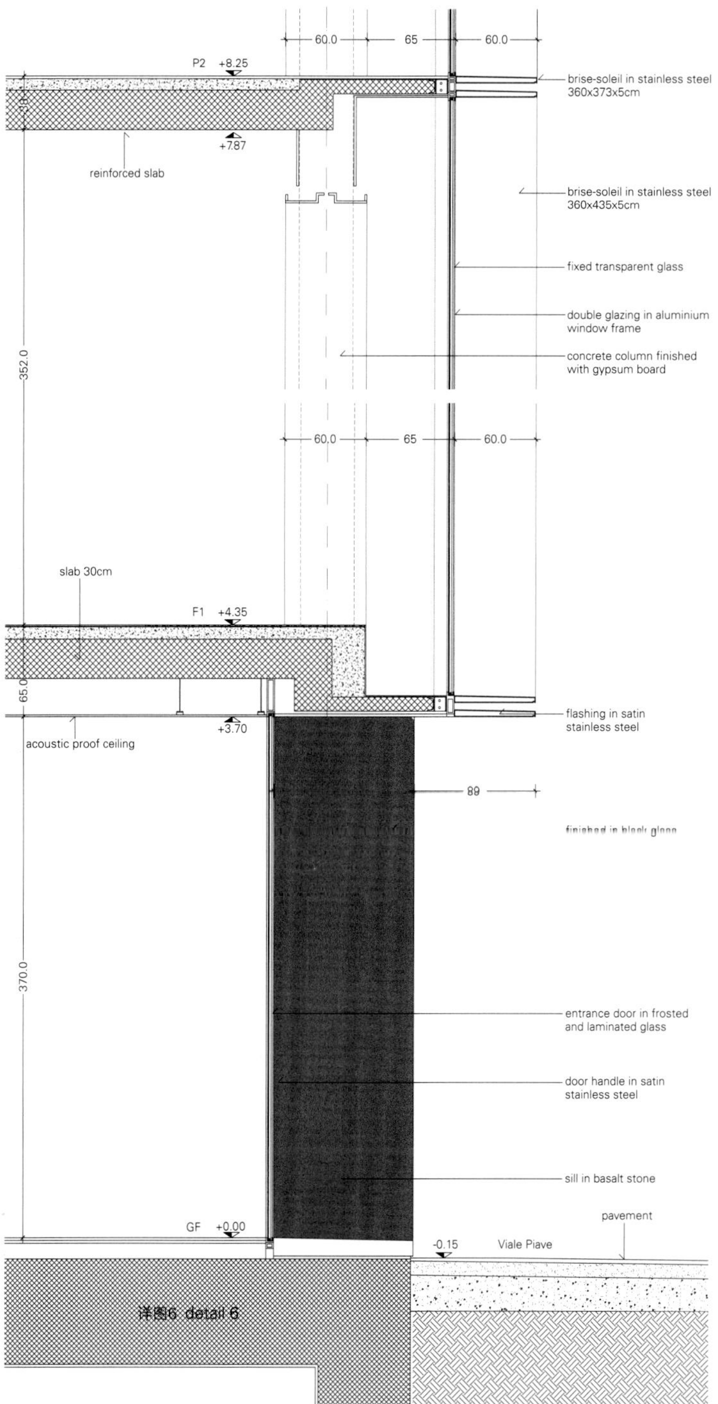

详图6 detail 6

项目名称：Dolce&Gabbana Office / 地点：Viale Piave 24, Milan, Italy
建筑师：Francesco Fresa, Germán Fuenmayor, Gino Garbellini, Monica Tricario
项目团队：Uca Lazzerotti, Micaela Bonomessi, Silvia Calzetti, Laura Cassani, Gabriele Coi, Marco Dragoni, Davide Fascione, Suewoo Kim, Magali Roig Liverato, Michele Megna, Elena Migliorati, Gianni Mollo, Andrea Palaia, Salvatore Seggio
机电工程师：GTEC a.s
照明工程师：Rossi Bianchi Lighting Design
客户：Dolce&Gabbana s.r.l.
建筑面积：8,500m² / 总建筑面积：10,000m²
设计时间：2009 / 竣工时间：2012
摄影师：©Andrea Martiradonna (courtesy of the architect)

©Enrico Basili

>>174

Piuarch

Francesco Fresa, Germán Fuenmayor, Gino Garbellini and Monica Tricario formed the Piuarch studio in 1996. Located in Brera, centre of Milan, it has grown to become a group of more than forty architects and engineers since 2008. Its research has been concentrated on architectural quality, with particular regard to the use of natural materials and natural light, the dialogue between architecture and its context, and the relationship between architecture and art. The encounter with Dolce&Gabbana marked the beginning of a collaboration that has been going on for ten years and that has resulted in the realization of more than forty boutiques all over the world, the Milan headquarters of Dolce&Gabbana and D&G, the Metropol theatre and the Factory in Incisa in Val d'Arno.

>>60

Skidmore, Owings & Merrill LLP

Skidmore, Owings & Merrill LLP (SOM) is one of the largest and leading architecture, interior design, engineering, and urban-planning companies in the world. It was formed in Chicago in 1936 by Louis Skidmore and Nathaniel Owings; in 1939 they were joined by John O. Merrill. Their first branch was opened in New York, 1937. The American Institute of Architects (AIA) has recognised SOM twice with its highest honour, the Architecture Company Award in 1962 and again in 1996. Their offices are located in London, Chicago, New York, San Francisco, Los Angeles, Washington, DC, Hong Kong, Shanghai, and Abu Dhabi.

>>40

Ateliers Jean Nouvel

Jean Nouvel was born in Fumel, France in 1945. Started his first architecture practice in 1970 and Obtained his degree at ENSBA(Ecole Nationale Supérieure des Beaux-Arts), Paris in 1972. In 1989, The Arab World Institute in Paris was awarded the Aga-Khan Prize. In 2000, he received the Lion d'Or of the Venice Biennale.

In 2001, he received the Royal Gold Medal of the RIBA, the Praemium Imperial of Japan's Fine Arts Association and the Borromini Prize given to architects under 40. Was appointed Docteur Honoris Causa of the Royal College of Art in London en 2002. Also received the International Highrise Award 2006 and prestigious Pritzker Prize in 2008. In France, he received many prizes including the Gold Medal of the French Academy of Architecture, the Équerre d'Argent and National Grand Prize for Architecture.

Isabel Potworowski

Has graduated from TU Delft with a Master in Architecture, and currently works for Barcode Architects in Rotterdam. During her graduate studies, she was a member of the editorial committee and wrote several articles for the independent student journal "Pantheon". Originally from Canada, completed her Bachelor in Architecture at McGill University in Montreal, where she was awarded the Louis Robertson book prize for the highest grade. Besides, studied for one semester at the Politecnico di Milano. Has worked at ONPA Architects and Manasc Isaac Architects, both in Edmonton, Canada.

Fabrizio Aimar

Graduated cum laude from Politecnico di Torino(Italy). Worked for five years for a civil and infrastructural engineering office in Torino. During this period, he has developed different structural projects in collaboration with some international architectural firms such as Jean Nouvel, Renzo Piano, Mario Cucinella and Aymeric Zublena. In 2014, Fabrizio has worked on the new Intesa Sanpaolo skyscraper in Torino designed by Renzo Piano(RPBW), and in the same year he has founded his own practice. Since 2010, he has been a contributor for the Italian architectural magazine "Il Giornale dell'Architettura" and also a member of Cultural Committee of the Asti local board of Architects. Since 2014, he is an external writer for the Italian technical websites "architetto.info" and "ingegneri.info".

>>24

CRAB Studio

The Cook Robotham Architectural Bureau was established in 2006 by directors Professor Sir Peter Cook[right] and Gavin Robotham[left]. Peter Cook(1936) was born in Southend-on Sea and studied at Art University Bournemouth and AA School in London. Was knighted by the Queen in 2007. Is a Royal Academician and a Commander of the Order of Arts and Letters and a Senior Fellow of the Royal College of Art, London. Gavin Robotham(1969) was born in North Walsham and studied at Bartlett School of Architecture and Harvard Graduate School of Design. Is director and design team leader of CRAB Studio and continues to lecture design and acts as a consultant.

©Miquel Tres

>>78

:studio Carme Pinós

;arme Pinós was born in 1954 and graduated from the ETSAB in Barcelona in 1979. Set up her own studio in 1991 after winning international ecognition for her work with Enric Miralles in projects such as the Igualada Cemetery. Has taught at the several higher education institutions, ıcluding the University of Illinois at Urbana-Champaign, Columbia University, EPFL, ETSAB and at Harvard GSD, among others. Some of her ıost recent and acclaimed projects include the Departments Building of the Vienna University of Economics and Business and the Cube I and Towers in Guadalajara. Received the First Prize at the 2008 Spanish Biennial of Architecture. Has recently been honoured with the Creu de ant Jordi Medal by the Government of Catalonia. Made Honorary member of the AIA in 2011 and RIBA International Fellow in 2013.

>>90

:. F. Møller Architects

s one of Scandinavia's oldest and largest architectural practices. Was founded in 1924 by the now deceased Danish architect C. F. Møller. Today has more than 350 employees in Head office in Aarhus, Denmark and some branches in Copenhagen, Aalborg, Oslo, Stockholm and London.)ver the years, have won a large number of national and international competitions. Also, has been exhibited locally as well as internationally at laces like at RIBA in London, the Venice Biennale, and the Danish Cultural Institute in Beijing. In 2015, it was named one of the Top 10 Most ınovative Companies in Architecture.

>>130

acques Ferrier Architecture

acques Ferrier studied at the National Superior school of Architecture(ENSA) Paris-Belleville and the Central school of Paris. Created his own rchitecture firm in Paris in 1993 and has since worked both in France and on the international stage. His portfolio of work includes cultural facil-:ies, showcase buildings, public buildings and research centres, as well as urban development projects that chime with the ethos and philoso-hy of the agency: to create architecture and cities that contribute to a sustainable society. Has been nominated three times for the Grand Prix lational d'Architecture and made both a Chevalier de l'Ordre National du Mérite and a Chevalier des Arts et des Lettres.

>>112

ιAU architects

Vas founded by German architect, Thomas Rau(1960) in 1992 at Amsterdam. Graduating from the University of Bonn, Thomas Rau studied rchitecture at the Alanus University of Arts and Social Sciences and RWTH Aachen University. He also founded One Planet Architecture insti-ıte (OPAi) in 2008 and Turntoo, a service company in 2010. The RAU architectural team designs buildings with a strong emphasis on sustain-

图书在版编目(CIP)数据

创意办公 : 汉英对照 / 法国让・努维尔建筑师事务所等编 ; 孙探春等译. — 大连 : 大连理工大学出版社, 2017.3

(建筑立场系列丛书)

ISBN 978-7-5685-0747-9

Ⅰ. ①创… Ⅱ. ①法… ②孙… Ⅲ. ①办公建筑—建筑设计—汉、英 Ⅳ. ①TU243

中国版本图书馆CIP数据核字(2017)第055590号

出版发行:大连理工大学出版社
（地址:大连市软件园路80号　　邮编:116023）
印　　刷:上海锦良印刷厂
幅面尺寸:225mm×300mm
印　　张:11.75
出版时间:2017年3月第1版
印刷时间:2017年3月第1次印刷
出 版 人:金英伟
统　　筹:房　磊
责任编辑:张昕焱
封面设计:王志峰
责任校对:张媛媛
书　　号:978-7-5685-0747-9
定　　价:258.00元

发　行:0411-84708842
传　真:0411-84701466
E-mail:12282980@qq.com
URL: http://www.dutp.cn

本书如有印装质量问题,请与我社发行部联系更换。

>>142

Cuantics Creatives Lab

Diego Soroa is a multidisciplinary creative, architect, researcher and concept designer who was born in Bilbao, 1975. Has been leading Cuantics Creative Lab for applied creativity since 2006 and obtained Best Entrepreneur Award from the Bilbao Town Hall. Works as a creative consultant for diverse institutions promoting creativity and providing new concepts, tools and perspectives for social, organizational and technological innovation. As an architect he is mainly focused on re-occupation of public spaces. Is also a creative coach and a lecturer on creativity, design thinking and futurization.

>>158

Office AT

Was founded in 2002 by Surachai Akekapobyotin and Juthathip Techacumreon, both B.Arch at the KMIT(King Mongkut's Institute of Technology) and M.Arch at the University of Michigan.
Director, Juthathip Techachumreon has worked at the Meridian Design Associates Architects, New York and taught at the ABAC(Assumption University of Thailand) and KMIT. Managing Director, Surachai Akekapobyotin was a visiting lecturer of the ABAC. Worked at the ASA(Association of Siamese Architects Under Royal Patronage), Plan Associates(Bangkok) and Meltzer/Mandl Architects(New York). Won several international and Thailand competitions such as ASA Gold Medal award, Silver Medal award and Citation Award.

>>130

>>112